SpringerBriefs in Optimization

SpringerBriefs present concise summaries of cutting-edge research and practical applications across a wide spectrum of fields. Featuring compact volumes of 50 to 125 pages, the series covers a range of content from professional to academic. Briefs are characterized by fast, global electronic dissemination, standard publishing contracts, standardized manuscript preparation and formatting guidelines, and expedited production schedules.

Typical topics might include:

- A timely report of state-of-the art techniques
- A bridge between new research results, as published in journal articles, and a contextual literature review
- A snapshot of a hot or emerging topic
- An in-depth case study
- A presentation of core concepts that students must understand in order to make independent contributions

SpringerBriefs in Optimization showcase algorithmic and theoretical techniques, case studies, and applications within the broad-based field of optimization. Manuscripts related to the ever-growing applications of optimization in applied mathematics, engineering, medicine, economics, and other applied sciences are encouraged.

Titles from this series are indexed by Web of Science, Mathematical Reviews, and zbMATH.

Alexander J. Zaslavski

The Krasnoselskii-Mann Method for Common Fixed Point Problems

 Springer

Alexander J. Zaslavski
Department of Mathematics
Technion – Israel Institute of Technology
Haifa, Haifa, Israel

ISSN 2190-8354 ISSN 2191-575X (electronic)
SpringerBriefs in Optimization
ISBN 978-3-031-85840-6 ISBN 978-3-031-85841-3 (eBook)
https://doi.org/10.1007/978-3-031-85841-3

This Springer imprint is published by the registered company Springer Nature Switzerland AG
The registered company address is: Gewerbestrasse 11, 6330 Cham, Switzerland

If disposing of this product, please recycle the paper.

Preface

In this book we study the Krasnoselskii-Mann method, which is of great importance and interest, in order to solve a common fixed point problem with a finite family of quasi-nonexpansive mappings acting of a convex set in a metric space with a hyperbolic structure. We consider several versions of the Krasnoselskii-Mann method associated with iterative algorithms, the Cimmino algorithm, methods with remotest set control, and dynamic string-averaging algorithms. Our goal is to study the convergence of iterates and obtain approximate solutions of the common fixed point problem under the presence of computational errors. We show that our methods generate a good approximate solution, if the sequence of computational errors is bounded from above by a constant. Moreover, for a known computational error, we find out what an approximate solution can be obtained and how many iterates one needs for this.

The monograph contains six chapters. Chapter 1 is an introduction. In Chap. 2 we study the convergence of iterative methods for solving common fixed point problems in a W-hyperbolic space using Krasnoselskii-Mann iterations acting in a W-space $X = (X, d, w)$, where (X, d) is a metric space and $W : X \times X \times [0, 1] \to X$ satisfies certain assumptions. We think of $W(x, y, \lambda)$ as a convex combination of the points $x, y \in X$ with the coefficients $1 - \lambda$, λ. We also assume that the W-hyperbolic space X has a structure (X, η), where $\eta : (0, \infty) \times (0, 2] \to (0, 1]$ is a so-called modulus of uniform convexity. In this chapter we study the behavior of exact and inexact iterates of this algorithm considering summable and nonsummable computational errors. In Chap. 3 we study the same common fixed point problem using an algorithm with remotest set control, and in Chap. 4 we use Krasnoselskii-Mann iterations for solving set-valued inclusions in the W-hyperbolic space. Our main goal is to obtain an approximate solution of the problem in the presence of computational errors.

In Chap. 5 we study the convergence of the Cimmino algorithm for solving the common fixed point problems in a normed space using Krasnoselskii-Mann iterations.

In Chap. 6 we study the convergence of dynamic string-averaging methods for solving the common fixed point problems in a normed space considered in Chap. 5.

Our main goal is to obtain an approximate solution of the problem in the presence of computational errors. It should be mentioned that string-averaging algorithms were first introduced by Y. Censor, T. Elfving, and G. T. Herman in [24] for solving a convex feasibility problem, when a given collection of sets is divided into blocks and the algorithms operate in such a manner that all the blocks are processed in parallel. Iterative methods for solving common fixed point problems is a special case of dynamic string-averaging methods with only one block. Iterative methods, the Cimmono algorithm, and dynamic string-averaging methods are important tools for solving common fixed point problems in a Hilbert space [23–27, 29, 31, 36, 37, 85, 98, 111, 142, 145, 149].

Rishon LeZion, Israel   Alexander J. Zaslavski
June 2024

Contents

Chapter 1
Introduction

Abstract In this book we study the Krasnoselskii-Mann method in order to solve a common fixed point problem with a finite family of quasi-nonexpansive mappings acting of a convex set in a metric space with a hyperbolic structure. We consider several versions of the Krasnoselskii-Mann method associated with iterative algorithms, the Cimmino algorithm, methods with remotest set control, and dynamic string-averaging algorithms. Our goal is to study the convergence of iterates and obtain approximate solutions of the common fixed point problem under the presence of computational errors.

1.1 Notation

The fixed point theory of nonlinear operators has been a rapidly growing area of research. The starting point of this theory is Banach's classical theorem [7] concerning the existence of a unique fixed point for a strict contraction. The main goals on this theory are to show the existence of a fixed point for a given nonlinear mapping and to construct an iterative process which generates approximate fixed point and in some cases converge to a fixed point of the mapping [1, 51, 52, 61–66, 76, 105, 110, 114, 117, 131, 133, 134, 136, 138, 149, 152, 153]. The fixed point theory contains the study of various classes of nonlinear single-valued [4, 8, 27, 28, 42, 47, 65, 66, 69–71, 82, 83, 89, 90, 92, 99, 102, 103, 107–109] as well as set-valued mappings [45, 88, 91, 96, 104, 106, 115, 139, 155] including nonexpanive mappings [9, 13, 25, 29, 32–35, 39, 40, 72, 74, 75, 78–81, 85, 100, 101, 111, 113, 125, 132, 143, 146, 154], monotone nonexpansive mappings [46, 116, 118, 119, 123], locally nonexpansive mappings [121, 122, 124], and mappings in metric spaces with a graph [59, 97, 120, 130]. It plays an important role in the study of variational inequalities [11, 15–17, 19–21, 38, 41, 49, 50, 54–57, 60, 67, 73, 128, 129, 135, 140], monotone operators [10, 126, 150, 151], feasibility problems [14, 22, 24, 26, 36, 37, 48, 84, 93–95, 98, 112, 137, 141, 142, 145, 148], optimization theory [6, 11, 31, 127], proximal point methods [3, 5, 18, 43, 58, 87, 144, 150, 151], superiorization [23, 30], and their applications in engineering, medical, and the natural sciences.

© The Author(s), under exclusive license to Springer Nature Switzerland AG 2025   1
A. J. Zaslavski, *The Krasnoselskii-Mann Method for Common Fixed
Point Problems*, SpringerBriefs in Optimization,
https://doi.org/10.1007/978-3-031-85841-3_1

In this book we study the Krasnoselskii-Mann method, which is of great importance and interest [12, 44, 53, 68, 77, 86, 147], in order to solve a common fixed point problem with a finite family of quasi-nonexpansive mappings acting of a convex set in a metric space with a hyperbolic structure. We consider several versions of the Krasnoselskii-Mann method associated with iterative algorithms, the Cimmino algorithm, methods with remotest set control, and dynamic string-averaging algorithms. Our goal is to study the convergence of iterates and obtain approximate solutions of the common fixed point problem under the presence of computational errors.

In our book we use the following notation. Let (X, d) be a metric space. For each $x \in X$ and each set $E \subset X$, put

$$d(x, E) = \inf\{d(x, y) : y \in E\}.$$

Note that $d(x, \emptyset) = \infty$, $x \in X$. For every pair of non-empty sets $A, B \subset X$, put

$$H(A, B) = \max\{\sup\{d(x, B) : x \in A\}, \ \sup\{d(y, A) : y \in B\}\}.$$

Given a mapping $P : X \to X$, we define $P^0 = I$, the identity self-mapping of X, $P^1 = P$, $P^{i+1} = P \circ P^i$ for all nonnegative integers i, and set

$$\text{Fix}(P) = \{z \in X : P(z) = z\}.$$

For each $x \in X$ and each $r > 0$, set

$$B_d(x, r) = \{y \in X : d(x, y) \le r\}.$$

If the metric d is understood, then

$$B(x, r) = B_d(x, r), \ x \in X, \ r > 0.$$

For each $x \in R^1$, set

$$\lfloor x \rfloor = \max\{i : i \text{ is an integer and } i \le x\}.$$

Denote by $\text{Card}(A)$ the cardinality of a set A. We assume that the sum over empty set is zero and that the infimum of an empty set is ∞. The set $X \times X$ is equipped with the metric

$$d_1((x_1, x_2), (y_1, y_2)) = d(x_1, y_1) + d(x_2, y_2), \ x_i, y_i \in X, \ i = 1, 2.$$

For each set $A \subset X$ denote by $\text{cl}(A)$ its closure.

1.2 Krasnoselskii-Mann Iterations

In Chap. 2 we study the convergence of iterative methods for solving common fixed point problems in a W-hyperbolic space using Krasnoselskii-Mann iterations acting in a W-space $X = (X, d, w)$, where (X, d) is a metric space and $W : X \times X \times [0, 1] \to X$ satisfies certain assumptions which are introduced and discussed in Sect. 2.1. We think of $W(x, y, \lambda)$ as convex combinations of the points $x, y \in X$ with the coefficients $1 - \lambda, \lambda$ and use the notation

$$W(x, y, \lambda) := (1 - \lambda)x \oplus \lambda y, \ x, y \in X, \ \lambda \in [0, 1].$$

We also assume that the W-hyperbolic space X has a structure (X, η), where $\eta : (0, \infty) \times (0, 2] \to (0, 1]$ is a so-called modulus of uniform convexity which is stated in Sect. 2.1. Note that convex subsets of normed spaces are W-hyperbolic spaces. Other examples of W-hyperbolic spaces are hyperbolic spaces [114], Busemann spaces [4], and $CAT(0)$ spaces [71].

A non-empty set $C \subset X$ is convex if for each $x, y \in C$ and each $\lambda \in [0, 1]$,

$$(1 - \lambda)x \oplus \lambda y \in C.$$

Assume that m is a natural number and that for $i = 1, \ldots, m$ and $T_i : C \to C$. Set

$$\text{Fix}(T_i) := \{z \in C : T_i(z) = z\}, \ i = 1, \ldots, m$$

and set

$$F = \cap_{i=1}^{m} \text{Fix}(T_i).$$

For every $\epsilon > 0$ and every $i \in \{1, \ldots, m\}$, put

$$F_\epsilon(T_i) = \{x \in C : d(x, T_i(x)) \leq \epsilon\},$$

$$\tilde{F}_\epsilon(T_i) = \{y \in C : d(y, F_\epsilon(T_i)) \leq \epsilon\},$$

$$F_\epsilon = \cap_{i=1}^{m} F_\epsilon(T_i),$$

and

$$\tilde{F}_\epsilon = \cap_{i=1}^{m} \tilde{F}_\epsilon(T_i).$$

A point belonging to the set F is a solution of our common fixed point problem, while a point which belongs to the set $\tilde{F}_\epsilon$ is its ϵ-approximate solution.
Fix $\theta \in C$.

Assume that $\bar{N} \geq m$ is a natural number. Denote by $\mathcal{R}$ the set of all mappings $r : \{0, 1, \ldots\} \to \{1, \ldots, m\}$ such that for each integer $k \geq 0$,

$$\{1, \ldots, m\} \subset \{r(i) : \; i \in \{k, \ldots, k + \bar{N} - 1\}\}.$$

Assume that $M_0 > 0$ and the following assumption holds.

(A) For each $\delta \in (0, 1)$ there exists $p_\delta \in B(\theta, M_0) \cap C$ such that for each $i \in \{1, \ldots, m\}$ and each $x \in C$,

$$d(T_i(x), p_\delta) \leq d(x, p_\delta) + \delta.$$

Now we describe our algorithm.
Initialization: Choose

$$x \in C \cap B(\theta, M_0), \; r \in \mathcal{R}$$

and set $x_0 = x$.

Iterative step: For each integer $i \geq 0$ choose $\alpha_i \in (0, 1)$ and set

$$x_{i+1} = (1 - \alpha_i)x_i \oplus \alpha_i T_{r(i)}(x_i).$$

In Chap. 2 we study the behavior of exact and inexact iterates of this algorithm considering summable and nonsummable computational errors.

In Chap. 3 we study the same common fixed point problem using an algorithm with remotest set control which is described below.

Initialization: Choose

$$x \in C \cap B(\theta, M_0)$$

and set $x_0 = x$.

Iterative step: For each integer $i \geq 0$ choose $\alpha_i \in (0, 1)$ and $r(i) \in \{1, \ldots, m\}$ such that

$$d(x_i, T_{r(i)}(x_i)) = \max\{d(x_i, T_s(x_i)) : \; s = 1, \ldots, m\}$$

and set $x_{i+1} = (1 - \alpha_i)x_i \oplus \alpha_i T_{r(i)}(x_i)$.

In Chap. 4 we use Krasnoselskii-Mann iterations for solving set-valued inclusions in the W-hyperbolic space. Our main goal is to obtain an approximate solution of the problem in the presence of computational errors.

We assume that $T : C \to 2^C \setminus \{\emptyset\}$, $T(x)$ is bounded for each $x \in X$ and the following assumption holds.

(B) For each $\delta \in (0, 1)$ there exists $p_\delta \in B(\theta, M_0) \cap C$ such that for each $x \in C$ and each $y \in T(x)$,

$$d(y, p_\delta) \leq d(x, p_\delta) + \delta.$$

Now we describe our algorithm.
Initialization: Let $\delta > 0$. Choose

$$x \in C \cap B(\theta, M_0)$$

and set

$$x_0 = x.$$

Iterative step: For each integer $i \geq 0$ choose $\alpha_i \in (0, 1)$ and $y_i \in T(x_i)$ such that

$$d(x_i, y_i) \geq \sup\{d(x_i, y) : y \in T(x_i)\} - \delta$$

and set

$$x_{i+1} = (1 - \alpha_i)x_i \oplus \alpha_i y_i.$$

In Chap. 5 we study the convergence of the Cimmino algorithm for solving the common fixed point problems in a normed space using Krasnoselskii-Mann iterations. Our main goal is to obtain an approximate solution of the problem in the presence of computational errors. We show that the method generates a good approximate solution, if the sequence of computational errors is bounded from above by a constant. Moreover, for a known computational error, we find out what an approximate solution can be obtained and how many iterates one needs for this. We assume that the W-hyperbolic space $X = (X, d, W)$ equipped with the structure (X, η), where $\eta : (0, \infty) \times (0, 2] \to (0, 1]$ is the modulus of uniform convexity, is also a normed space $(X, \| \cdot \|)$ and that

$$d(x, y) = \|x - y\|, \; x, y \in X$$

and that for all $\alpha \in [0, 1]$, $x, y \in X$,

$$(1 - \alpha)x \oplus \alpha y = (1 - \alpha)x + \alpha y.$$

Assume that $C \subset X$ is a non-empty convex set, m is a natural number and that for $i = 1, \ldots, m$ and $T_i : C \to C$. Assume that

$$\widehat{\Delta} \in (0, , m^{-1}],$$

$M_0 > 0$, and assumption (A) holds.

Now we describe our algorithm.

Initialization: Choose $x \in C \cap B(0, M_0)$, and set $x_0 = x$.

Iterative step: For each integer $i \geq 0$ choose $\alpha_i \in (0, 1)$ and $\beta_{i,j} \geq \widehat{\Delta}$, $j = 1, \ldots, m$ such that

$$\sum_{j=1}^{m} \beta_{i,j} = 1$$

and set

$$x_{i+1} = (1 - \alpha_i)x_i + \alpha_i \sum_{j=1}^{m} \beta_{i,j} T_j(x_i).$$

In Chap. 6 we study the convergence of dynamic string-averaging methods for solving the common fixed point problems in a normed space considered in Chap. 5. Our main goal is to obtain an approximate solution of the problem in the presence of computational errors. It should be mentioned that string-averaging algorithms were first introduced by Y. Censor, T. Elfving, and G. T. Herman in [24] for solving a convex feasibility problem, when a given collection of sets is divided into blocks and the algorithms operate in such a manner that all the blocks are processed in parallel. Iterative methods for solving common fixed point problems is a special case of dynamic string-averaging methods with only one block. Iterative methods, the Cimmono algorithm, and dynamic string-averaging methods are important tools for solving common fixed point problems in a Hilbert space [23–27, 29, 31, 36, 37, 85, 98, 111, 142, 145, 149].

Chapter 2
Iterative Methods

Abstract In this chapter we study the convergence of iterative methods for solving common fixed point problems in a W-hyperbolic space using Krasnoselskii-Mann iterations. Our main goal is to obtain an approximate solution of the problem in the presence of computational errors. We show that the iterative method generates a good approximate solution, if the sequence of computational errors is bounded from above by a constant. Moreover, for a known computational error, we find out what an approximate solution can be obtained and how many iterates one needs for this.

2.1 Preliminaries

A W-space is a structure (X, d, W) where (X, d) is a metric space and $W : X \times X \times [0, 1] \to X$. We think of $W(x, y, \lambda)$ as convex combinations of the points $x, y \in X$ with the coefficients $1 - \lambda$, λ and use the notation

$$W(x, y, \lambda) := (1 - \lambda)x \oplus \lambda y, \ x, y \in X, \ \lambda \in [0, 1].$$

A W-space (X, d, W) is called a W-hyperbolic space [70] if for all $x, y, w, z \in X$ and all $\lambda, \tilde{\lambda} \in [0, 1]$ the following properties hold:

(W1) $d(z, (1 - \lambda)x \oplus \lambda y) \leq (1 - \lambda)d(z, x) + \lambda d(z, y)$.
(W2) $d((1 - \lambda)x \oplus \lambda y, (1 - \tilde{\lambda})x \oplus \tilde{\lambda}y) \leq |\lambda - \tilde{\lambda}|d(x, y)$.
(W3) $(1 - \lambda)x \oplus \lambda y = \lambda y \oplus (1 - \lambda)x$.
(W4) $d((1 - \lambda)x \oplus \lambda z, (1 - \lambda)y \oplus \lambda w) \leq (1 - \lambda)d(x, y) + \lambda d(z, w)$.

It is obvious that convex subsets of normed spaces are W-hyperbolic spaces. Other examples of W-hyperbolic spaces are hyperbolic spaces [114], Busemann spaces [4], and $CAT(0)$ spaces [71].

In the sequel we assume that (X, d, W) is a W-hyperbolic space which is denoted by X for simplicity.

© The Author(s), under exclusive license to Springer Nature Switzerland AG 2025
A. J. Zaslavski, *The Krasnoselskii-Mann Method for Common Fixed Point Problems*, SpringerBriefs in Optimization, https://doi.org/10.1007/978-3-031-85841-3_2

A non-empty set $C \subset X$ is convex if for each $x, y \in C$ and each $\lambda \in [0, 1]$,

$$(1 - \lambda)x \oplus \lambda y \in C.$$

Recall that for each $x \in X$ and each set $E \subset X$,

$$d(x, E) = \inf\{d(x, y) : y \in E\}$$

and that for each $x \in X$ and each $r > 0$,

$$B(x, r) = \{y \in X : d(x, y) \le r\}.$$

For each $x \in X$, each non-empty set $E \subset X$, and each $\lambda \in [0, 1]$, set

$$\lambda x \oplus (1 - \lambda)E = \{\lambda x \oplus (1 - \lambda)y : y \in E\}.$$

The following result is well known in the literature [33, 132].

Proposition 2.1 *Let $x, y, w, z \in X$ and let $\lambda, \tilde{\lambda} \in [0, 1]$. Then*

$$d(x, (1 - \lambda)x \oplus \lambda y) \le \lambda d(x, y), \ \ d(y, (1 - \lambda)x \oplus \lambda y) \le (1 - \lambda)d(x, y) \quad (2.1)$$

and

$$d((1 - \lambda)x \oplus \lambda z, (1 - \tilde{\lambda})y \oplus \tilde{\lambda}w)$$

$$\le (1 - \lambda)d(x, y) + \lambda d(z, w) + |\lambda - \tilde{\lambda}|d(y, w). \quad (2.2)$$

Clearly, Eq. (2.1) follows from (W1) and properties (W4) and (W2) imply (2.2).

Let $X = (X, d, W)$ be a W-hyperbolic space. We assume that the W-hyperbolic space X has a structure (X, η), where $\eta : (0, \infty) \times (0, 2] \to (0, 1]$ is a so-called modulus of uniform convexity such that the following assumptions hold:

(B1)   For each $r > 0$, each $\epsilon \in (0, 2]$, and each $x, y, a \in X$, if

$$d(x, a), \ d(y, a) \le r \text{ and } d(x, y) \ge \epsilon r,$$

then

$$d(2^{-1}x + 2^{-1}y, a) \le (1 - \eta(r, \epsilon))r.$$

(B2)   For each $\epsilon \in (0, 2]$ and each numbers s, r satisfying $0 < r \le s$,

$$\eta(s, \epsilon) \le \eta(r, s).$$

It turns out that this class of spaces is an appropriate setting for obtaining quantitative results on the asymptotic behavior of the Mann iteration for nonexpansive mappings (see [72, 78, 79]), as well as of the Picard iteration for firmly nonexpansive mappings [4]. It contains uniformly convex normed spaces and CAT(0) space.

In this chapter we use the following lemma. For its proof see Lemma 2.1 (iv) of [79].

Lemma 2.2 *Let* $r > 0$, $\epsilon \in (0, 2]$, $x, y, a \in X$,

$$d(x, a) \leq r, \ d(y, a) \leq r, \ d(x, y) \geq \epsilon r.$$

Then for each $\lambda \in [0, 1]$ *and each* $s \geq r$,

$$d((1 - \lambda)x \oplus \lambda y, a) \leq (1 - 2\lambda(1 - \lambda)\eta(s, \epsilon))r.$$

In this chapter we consider a common fixed point problem with a finite family of quasi-nonexpansive self-mappings of a convex subset of the metric space X with the structure (X, η). The results of the chapter are new.

2.2 A Common Fixed Point Problem

We continue to use the notation and definitions and assume the assumptions introduced in Sect. 2.1. In particular, $X = (X, d, W)$ is a W-hyperbolic space equipped with the structure (X, η), where $\eta : (0, \infty) \times (0, 2] \to (0, 1]$ is the modulus of uniform convexity. Let $C \subset X$ be a non-empty convex set.

Assume that m is a natural number and that for $i = 1, \ldots, m$ and $T_i : C \to C$. Set

$$\text{Fix}(T_i) := \{z \in C : T_i(z) = z\}, \ i = 1, \ldots, m$$

and set

$$F = \cap_{i=1}^m \text{Fix}(T_i).$$

For every $\epsilon > 0$ and every $i \in \{1, \ldots, m\}$ put

$$F_\epsilon(T_i) = \{x \in C : d(x, T_i(x)) \leq \epsilon\},$$

$$\tilde{F}_\epsilon(T_i) = \{y \in C : d(y, F_\epsilon(T_i)) \leq \epsilon\},$$

$$F_\epsilon = \cap_{i=1}^m F_\epsilon(T_i)$$

and

$$\tilde{F}_\epsilon = \cap_{i=1}^m \tilde{F}_\epsilon(T_i).$$

A point belonging to the set F is a solution of our common fixed point problem, while a point which belongs to the set $\tilde{F}_\epsilon$ is its ϵ-approximate solution.

Fix $\theta \in C$.

Proposition 2.3 *Assume that for every $i \in \{1, \ldots, m\}$, every $x \in C$, and every $y \in C$,*

$$d(T_i(x), T_i(y)) \le d(x, y)$$

and that $\epsilon > 0$. Then

$$\tilde{F}_\epsilon \subset F_{3\epsilon}.$$

Proof Let $z \in \tilde{F}_\epsilon$ and $i \in \{1, \ldots, m\}$. Then $d(z, F_\epsilon(T_i)) \le \epsilon$. Let $\delta > 0$. There exists

$$y \in F_\epsilon(T_i) \text{ such that } d(z, y) \le \epsilon + \delta.$$

Now we have $d(y, T_i(y)) \le \epsilon$,

$$d(z, T_i(x)) \le d(z, y) + d(y, T_i(y)) + d(T_i(y), T_i(z))$$

$$\le \epsilon + 2d(z, y) \le 3\epsilon + 3\delta.$$

Since δ is a positive number, we conclude that

$$d(z, T_i(z)) \le 3\epsilon, \ \ z \in F_{3\epsilon}(T_i) \text{ and } z \in F_{3\epsilon}.$$

Proposition 2.3 is proved.

2.3   The Basic Lemma

The following lemma is an important ingredient in our study.

Lemma 2.4 *Assume that $S : C \to C$, $M > 0$, $\alpha, \gamma \in (0, 1)$,*

$$0 < \delta \le 4^{-1}\gamma\alpha(1 - \alpha)\eta(2M + 1, \gamma(2M + 1)^{-1}), \tag{2.3}$$

$$p \in B(\theta, M) \cap C, \tag{2.4}$$

$$d(S(x), p) \leq d(x, p) + \delta, \, , \quad x \in C, \tag{2.5}$$

$$u \in B(\theta, M) \cap C, \tag{2.6}$$

$$v = (1 - \alpha)u \oplus \alpha S(u) \tag{2.7}$$

and that

$$d(u, S(u)) \geq \gamma.$$

Then the following assertions hold: 1.

$$d(v, p) \leq d(u, p) + 1 \leq 2M + 1,$$

$$d(v, p) \leq d(u, p) - 4^{-1}\gamma\alpha(1 - \alpha)\eta(2M + 1, \gamma(2M + 1)^{-1}).$$

2. Assume that

$$\delta_1 = 8^{-1}\gamma\alpha(1 - \alpha)\eta(2M + 1, \gamma(2M + 1)^{-1}), \, \widehat{v} \in C \cap B(v, \delta_1).$$

Then

$$d(\widehat{v}, p) \leq d(u, p) - 8^{-1}\gamma\alpha(1 - \alpha)\eta(2M + 1, \gamma(2M + 1)^{-1}).$$

Proof Proposition 2.1 and Eqs. (2.4)–(2.7) imply that

$$d(v, p) = d((1 - \alpha)u \oplus \alpha S(u), p)$$

$$\leq (1 - \alpha)d(u, p) + \alpha d(S(u), p) \leq (1 - \alpha)d(u, p) + \alpha d(u, p) + \alpha\delta$$

$$\leq d(u, p) + 1 \leq d(u, \theta) + d(\theta, p) + 1 \leq 2M + 1, \tag{2.8}$$

$$d(v, \theta) \leq d(v, p) + d(p, \theta) \leq 3M + 1.$$

By (2.3), (2.5) and (2.7),

$$\gamma \leq d(u, S(u)) \leq d(u, p) + d(p, S(u)) \leq d(u, p) + d(p, u) + \delta,$$

$$d(u, p) \geq 2^{-1}(\gamma - \delta) \geq 4^{-1}\gamma. \tag{2.9}$$

In view of (2.4), (2.6) and (2.8),

$$d(u, S(u)) \geq \gamma \geq \gamma(\delta + d(u, p))(2M + 1)^{-1}. \tag{2.10}$$

By (2.5),

$$d(p, S(u)) \leq \delta + d(p, u). \tag{2.11}$$

Equations (2.4), (2.6), (2.7), (2.10), (2.11), Lemma 2.2 applied with

$$x = u, \ y = S(u), \ a = p, \ r = \delta + d(u, p),$$

$$s = 2M + 1, \ \epsilon = \gamma(2M + 1)^{-1}, \ \lambda = \alpha$$

and Eqs. (2.3), (2.9) imply that

$$d(v, p) = d((1 - \alpha)u \oplus \alpha S(u), p)$$

$$\leq (1 - 2\alpha(1 - \alpha)\eta(2M + 1, \gamma(2M + 1)^{-1}))(\delta + d(u, p))$$

$$\leq d(u, p) + \delta - 2\alpha(1 - \alpha)\eta(2M + 1, \gamma(2M + 1)^{-1})d(u, p)$$

$$\leq d(u, p) + \delta - 2^{-1}\gamma\alpha(1 - \alpha)\eta(2M + 1, \gamma(2M + 1)^{-1})$$

$$\leq d(u, p) - 4^{-1}\gamma\alpha(1 - \alpha)\eta(2M + 1, \gamma(2M + 1)^{-1}).$$

Assertion 1 is proved. Assertion 2 follows from Assertion 1. Lemma 2.4 is proved.

2.4   Inexact Iterates

Assume that $\bar{N} \geq m$ is a natural number. Denote by $\mathcal{R}$ the set of all mappings $r : \{0, 1, \ldots\} \to \{1, \ldots, m\}$ such that for each integer $k \geq 0$,

$$\{1, \ldots, m\} \subset \{r(i) : i \in \{k, \ldots, k + \bar{N} - 1\}\}.$$

Assume that $M_0 > 0$ and the following assumption holds:

(A)   For each $\delta \in (0, 1)$ there exists $p_\delta \in B(\theta, M_0) \cap C$ such that for each $i \in \{1, \ldots, m\}$ and each $x \in C$,

$$d(T_i(x), p_\delta) \leq d(x, p_\delta) + \delta.$$

For each $\delta \in (0, 1)$ let p_δ be as guaranteed by (A). Now we describe our algorithm.

Initialization: Choose

$$x \in C \cap B(\theta, M_0), \ r \in \mathcal{R}$$

and set $x_0 = x$.

Iterative step: For each integer $i \geq 0$ choose $\alpha_i \in (0, 1)$ and set

$$x_{i+1} = (1 - \alpha_i)x_i \oplus \alpha_i T_{r(i)}(x_i).$$

In this section we consider inexact iterates of this algorithm under the presence of summable computational errors.

Proposition 2.5 *Assume that* $\{\Delta_i\}_{i=0}^{\infty} \subset [0, \infty)$,

$$\infty \geq \Delta \geq \sum_{i=0}^{\infty} \Delta_i, \tag{2.12}$$

$\{\alpha_i\}_{i=0}^{\infty} \subset (0, 1)$,

$$x_0 \in B(\theta, M_0) \cap C, \tag{2.13}$$

$r : \{0, 1, \ldots\} \to \{1, \ldots, m\}$ *and that for each integer* $n \geq 0$,

$$x_{n+1} \in C \cap B((1 - \alpha_n)x_n \oplus \alpha_n T_{r(n)}(x_n), \Delta_n). \tag{2.14}$$

Then for each integer $k \geq 0$ *and each* $\delta \in (0, 1)$,

$$d(x_{k+1}, p_\delta) \leq d(x_k, p_\delta) + \delta + \Delta_k$$

and for each integer $n \geq 0$ *and each* $j \in \{1, \ldots, m\}$,

$$d(x_k, \theta) \leq 3M_0 + \Delta, \ d(T^j(x_n), \theta) \leq 3M_0 + \Delta.$$

Proof Let $\delta \in (0, 1)$. Proposition 2.1, assumption (A), and Eqs. (2.13) and (2.14) imply that for each integer $k \geq 0$ and each $j \in \{1, \ldots, m\}$,

$$d(x_{k+1}, p_\delta) \leq d(x_{k+1}, (1 - \alpha_k)x_k \oplus \alpha_k T_{r(k)}(x_k))$$

$$+ d((1 - \alpha_k)x_k \oplus \alpha_k T_{r(k)}(x_k), p_\delta)$$

$$\leq \Delta_k + (1 - \alpha_k)d(x_k, p_\delta) + \alpha_k d(T_{r(k)}(x_k), p_\delta)$$

$$\leq \Delta_k + (1 - \alpha_k)d(x_k, p_\delta) + \alpha_k d(x_k, p_\delta) + \delta$$

$$\leq \Delta_k + \delta + d(x_k, p_\delta),$$

$$d(x_k, p_\delta) \leq d(x_0, p_\delta) + k\delta + \sum_{i=0}^{k} \Delta_i,$$

$$d(T^j(x_k), p_\delta) \le d(x_k, p_\delta) + \delta \le d(x_0, p_\delta) + (k+1)\delta + \sum_{i=0}^{k} \Delta_i,$$

$$d(x_k, \theta) \le d(x_k, p_\delta) + d(p_\delta, \theta)$$

$$\le d(x_0, p_\delta) + k\delta + \sum_{i=0}^{k} \Delta_i + d(p_\delta, \theta) \le 3M_0 + k\delta + \sum_{i=0}^{k} \Delta_i,$$

$$d(T^j(x_k), \theta) \le d(T^j(x_k), p_\delta) + d(p_\delta, \theta)$$

$$\le d(x_0, p_\delta) + (k+1)\delta + \sum_{i=0}^{k} \Delta_i + M_0 \le 3M_0 + (k+1)\delta + \sum_{i=0}^{k} \Delta_i.$$

Since δ is any element of the interval $(0, 1)$, Proposition 2.5 is proved.

The following theorem shows that our algorithm generates approximate solutions of the common fixed point problem under the presence of summable computational errors.

Theorem 2.6 *Assume that* $\{\Delta_i\}_{i=0}^{\infty} \subset [0, \infty)$,

$$\Delta = \sum_{i=0}^{\infty} \Delta_i < \infty, \tag{2.15}$$

$\{\alpha_i\}_{i=0}^{\infty} \subset (0, 1)$,

$$x_0 \in B(\theta, M_0) \cap C, \ r \in \mathcal{R}, \tag{2.16}$$

for each integer $n \ge 0$,

$$x_{n+1} \in C \cap B((1 - \alpha_n)x_n \oplus \alpha_n T_{r(n)}(x_n), \Delta_n), \tag{2.17}$$

$\epsilon \in (0, 1)$, n_0 *and* Q *are natural numbers such that*

$$\Delta_i < (2\bar{N})^{-1}\epsilon \tag{2.18}$$

for each integer $i \ge n_0\bar{N}$ *and that*

$$\sum_{n=n_0}^{n_0+Q-1} \min\{\alpha_i(1 - \alpha_i) : \ i \in \{\bar{N}n, \ldots, (n+1)\bar{N} - 1\}\}$$

$$> 2\bar{N}(4M_0 + 2\Delta)\epsilon^{-1}\eta^{-1}(6M_0 + 2\Delta + 1, \epsilon(2\bar{N})^{-1}(6M_0 + 1 + 2\Delta)^{-1}). \tag{2.19}$$

Then there exists an integer $n \in \{n_0, \ldots, n_0 + Q - 1\}$ such that for each $i \in \{n\bar{N}, \ldots, (n+1)\bar{N} - 1\}$,

$$d(x_i, T_{r(j)}(x_i)) \leq \epsilon(2\bar{N})^{-1}, \; d(x_i, x_{i+1}) \leq \epsilon(2\bar{N})^{-1},$$

for each $i, j \in \{n\bar{N}, \ldots, (n+1)\bar{N}\}, d(x_i, x_j) \leq \epsilon$ and for each $i \in \{n\bar{N}, \ldots, (n+1)\bar{N}\}, x_i \in \tilde{F}_\epsilon$.

Proof Set

$$\epsilon_0 = \epsilon(2\bar{N})^{-1}. \tag{2.20}$$

Proposition 2.5 and (2.15)–(2.17) imply that for each integer $n \geq 0$,

$$d(x_n, \theta) \leq 3M_0 + \Delta, \; d(T_j(x_n), \theta) \leq 3M_0 + \Delta, \; j = 1, \ldots, m. \tag{2.21}$$

Fix a positive number δ such that

$$\delta < 4^{-1}\epsilon_0\eta(6M_0+2\Delta+1, \epsilon_0(6M_0+1+2\Delta)^{-1})\alpha_i(1-\alpha_i), \; i = 0, \ldots, (n_0+Q)\bar{N}. \tag{2.22}$$

Assume that a nonnegative integer $j \in \{0, \ldots, (n_0 + Q)\bar{N}\}$ and that

$$d(x_j, T_{r(j)}(x_j)) > \epsilon_0. \tag{2.23}$$

By assumption (A), Eqs. (2.21)–(2.23) and Lemma 2.4 applied with $S = T_{r(j)}$, $u = x_j, \alpha = \alpha_j, M = 3M_0 + \Delta, p = p_\delta, \gamma = \epsilon_0$, and

$$v = (1 - \alpha_j)x_j \oplus \alpha_j T_{r(j)}(x_j),$$

we have

$$d((1 - \alpha_j)x_j \oplus \alpha_j T_{r(j)}(x_j), p_\delta)$$

$$\leq d(x_j, p_\delta) - 4^{-1}\epsilon_0\alpha_j(1-\alpha_j)\eta(6M_0+2\Delta+1, \epsilon_0(6M_0+2\Delta+1)^{-1}). \tag{2.24}$$

In view of (2.17) and (2.24),

$$d(x_{j+1}, p_\delta) \leq d(x_{j+1}, (1 - \alpha_j)x_j \oplus \alpha_j T_{r(j)}(x_j))$$

$$+ d((1 - \alpha_j)x_j \oplus \alpha_j T_{r(j)}(x_j), p_\delta)$$

$$\leq d(x_j, p_\delta) - 4^{-1}\epsilon_0\alpha_j(1 - \alpha_j)\eta(6M_0 + 2\Delta + 1, \epsilon_0(6M_0 + 2\Delta + 1)^{-1}) + \Delta_j.$$

Thus we have shown that the following property holds:

(a) For each integer $j \in \{0, \ldots, (n_0 + Q)\bar{N}\}$ for which

$$d(x_j, T_{r(j)}(x_j)) > \epsilon_0$$

we have

$$d(x_{j+1}, p_\delta) \leq d(x_j, p_\delta)$$

$$- 4^{-1}\epsilon_0\alpha_j(1 - \alpha_j)\eta(6M_0 + 2\Delta + 1, \epsilon_0(6M_0 + 2\Delta + 1)^{-1}) + \Delta_j. \quad (2.25)$$

We show that there exists an integer $n \in \{n_0, \ldots, n_0 + Q - 1\}$ such that

$$d(x_i, T_{r(i)}(x_i)) \leq \epsilon_0, \;\; i \in \{n\bar{N}, \ldots, (n+1)\bar{N} - 1\}.$$

Assume the contrary. Then for each $n \in \{n_0, \ldots, n_0 + Q - 1\}$,

$$\max\{d(x_i, T_{r(i)}(x_i)) : \; i \in \{n\bar{N}, \ldots, (n+1)\bar{N} - 1\}\} > \epsilon_0. \quad (2.26)$$

Let $n \in \{n_0, \ldots, n_0 + Q - 1\}$. In view of (2.26), there exists $j \in \{n\bar{N}, \ldots, (n+1)\bar{N} - 1\}$ such that

$$d(x_j, T_{r(j)}(x_j)) > \epsilon_0.$$

Property (a) implies that Eq. (2.25) holds. Proposition 2.5 and (2.15)–(2.17), (2.25) imply that

$$d(x_{n\bar{N}}, p_\delta) - d(x_{(n+1)\bar{N}}, p_\delta)$$

$$= \sum_{i=n\bar{N}}^{(n+1)\bar{N}-1} (d(x_i, p_\delta) - d(x_{i+1}, p_\delta))$$

$$\geq \sum\{d(x_i, p_\delta) - d(x_{i+1}, p_\delta) : i \in \{n\bar{N}, \ldots, (n+1)\bar{N} - 1\} \setminus \{j\}\}$$

$$+ d(x_j, p_\delta) - d(x_{j+1}, p_\delta)$$

$$\geq -\delta(\bar{N} - 1) - \sum\{\Delta_i : \; i \in \{n\bar{N}, \ldots, (n+1)\bar{N} - 1\} \setminus \{j\}\}$$

$$+ 4^{-1}\epsilon_0\alpha_j(1 - \alpha_j)\eta(6M_0 + 2\Delta + 1, \epsilon_0(6M_0 + 2\Delta + 1)^{-1}) - \Delta_j$$

$$\geq -\delta\bar{N} - \sum_{i=n\bar{N}}^{(n+1)\bar{N}-1} \Delta_i$$

$$+ 4\epsilon_0 \eta (6M_0 + 2\Delta + 1, \epsilon_0 (6M_0 + 2\Delta + 1)^{-1}) \min\{\alpha_i (1 - \alpha_i):$$

$$i \in \{n\bar{N}, \ldots, (n+1)\bar{N} - 1\}\}. \tag{2.27}$$

By assumption (A), (2.21), and (2.27),

$$4M_0 + \Delta \geq d(x_{n_0\bar{N}}, \theta) + d(\theta, p_\delta) \geq d(x_{n_0\bar{N}}, p_\delta)$$

$$= d(x_{n_0\bar{N}}, p_\delta) - d(x_{(n_0+Q)\bar{N}}, p_\delta)$$

$$= \sum_{n=n_0}^{n_0+Q-1} (d(x_{n\bar{N}}, p_\delta) - d(x_{(n+1)\bar{N}}, p_\delta))$$

$$\geq -\delta\bar{N}Q - \sum_{i=n_0\bar{N}}^{(n_0+Q)\bar{N}-1} \Delta_i$$

$$+ (\sum_{i=n_0}^{n_0+Q-1} \min\{\alpha_i (1 - \alpha_i): \ i \in \{n\bar{N}, \ldots, (n+1)\bar{N} - 1\}\})$$

$$\times 4\epsilon_0 \eta (6M_0 + 2\Delta + 1, \epsilon_0 (6M_0 + 2\Delta + 1)^{-1}),$$

$$\sum_{i=n_0}^{n_0+Q-1} \min\{\alpha_i (1 - \alpha_i): \ i \in \{n\bar{N}, \ldots, (n+1)\bar{N} - 1\}\}$$

$$\leq 4^{-1}(4M_0 + 2\Delta + \delta\bar{N}Q)\epsilon_0^{-1}\eta^{-1}(6M_0 + 2\Delta + 1, \epsilon_0 (6M_0 + 2\Delta + 1)^{-1}).$$

Since δ is any positive number satisfying (2.22), the relation above contradicts our definition of Q (see (2.19) and (2.20)). The contradiction we have reached proves that there exists $n \in \{n_0, \ldots, n_0 + Q - 1\}$ such that

$$d(x_i, T_{r(i)}(x_i)) \leq \epsilon_0 = \epsilon(2\bar{N})^{-1}, \ i \in \{n\bar{N}, \ldots, (n+1)\bar{N} - 1\}. \tag{2.28}$$

Proposition 2.1 and Eqs. (2.17), (2.18), (2.20), and (2.28) imply that for each $i \in \{n\bar{N}, \ldots, (n+1)\bar{N} - 1\}$,

$$d(x_i, x_{i+1}) \leq d(x_i, (1 - \alpha_i)x_i \oplus \alpha_i T_{r(i)}(x_i))$$

$$+ d((1 - \alpha_i)x_i \oplus \alpha_i T_{r(i)}(x_i), x_{i+1})$$

$$\leq \alpha_i d(x_i, T_{r(i)}(x_i)) + \Delta_i \leq \epsilon_0 + \Delta_i \leq 2\epsilon_0 = \epsilon\bar{N}^{-1}. \tag{2.29}$$

In view if (2.29), for each $i, j \in \{n\bar{N}, \ldots, (n+1)\bar{N}\}$,

$$d(x_i, x_j) \leq 2\epsilon_0 \bar{N} = \epsilon. \tag{2.30}$$

Assume that $i \in \{n\bar{N}, \ldots, (n+1)\bar{N}\}$ and $s \in \{1, \ldots, m\}$. By (2.16) and (2.28), there exists $j \in \{n\bar{N}, \ldots, (n+1)\bar{N} - 1\}$ such that $s = r(j)$ and

$$d(x_j, T_s(x_j)) = d(x_j, T_{r(j)}(x_j)) \leq \epsilon_0.$$

Together with (2.30) this implies that $x_i \in \tilde{F}_\epsilon$ and completes the proof of Theorem 2.6.

Theorem 2.6 implies the following result.

Theorem 2.7 *Assume that* $\{\Delta_i\}_{i=0}^\infty \subset [0, \infty)$, $\sum_{i=0}^\infty \Delta_i < \infty$, $\{\alpha_i\}_{i=0}^\infty \subset (0, 1)$, $x_0 \in B(\theta, M_0) \cap C$, $r \in \mathcal{R}$, *for each integer* $n \geq 0$,

$$x_{n+1} \in C \cap B((1 - \alpha_n)x_n \oplus \alpha_n T_{r(n)}(x_n), \Delta_n)$$

and that

$$\sum_{n=0}^\infty \min\{\alpha_i(1 - \alpha_i) : i \in \{\bar{N}n, \ldots, (n+1)\bar{N} - 1\}\} = \infty.$$

Then there exists a strictly increasing sequence of natural numbers n_k *such that for each integer* $k \geq 1$, $x_{n_k} \in \tilde{F}_{1/k}$.

2.5   Exact Iterates

In this section we prove two theorems which describe the behavior of exact iterates of our algorithm.

Theorem 2.8 *Assume that* $\Lambda \in (0, 2^{-1})$,

$$\{\alpha_i\}_{i=0}^\infty \subset (\Lambda, 1 - \Lambda), \tag{2.31}$$

$$x_0 \in B(\theta, M_0) \cap C, \ r \in \mathcal{R}, \tag{2.32}$$

for each integer $n \geq 0$,

$$x_{n+1} = (1 - \alpha_n)x_n \oplus \alpha_n T_{r(n)}(x_n) \tag{2.33}$$

and $\epsilon \in (0, 1)$. Then

$$Card(\{n \in \{0, 1, \ldots\} : x_n \notin \tilde{F}_\epsilon\})$$

$$\leq 4\bar{N}^2(2M_0 + 1)\epsilon^{-1}\Lambda^{-2}\eta^{-1}(6M_0 + 1, \epsilon\bar{N}^{-1}(6M_0 + 1)^{-1}).$$

Proof Let Q be a natural number. Set

$$\epsilon_0 = \epsilon\bar{N}^{-1}. \tag{2.34}$$

Choose a positive number

$$\delta \leq 4^{-1}(Q + 1)^{-1}\epsilon_0\eta(6M_0 + 1, \epsilon_0(6M_0 + 1)^{-1})\alpha_i(1 - \alpha_i), \; i = 0, \ldots, Q. \tag{2.35}$$

Set

$$\Delta_0 = 4^{-1}\epsilon_0\Lambda^2\eta(6M_0 + 1, \epsilon_0(6M_0 + 1)^{-1}), \tag{2.36}$$

$$E_Q = \{n \in \{0, \ldots, Q\} : d(x_n, T_{r(n)}(x_n)) > \epsilon_0\}. \tag{2.37}$$

Assume that

$$n \in E_Q.$$

By assumption (A), Proposition 2.5, the relation above, Eqs. (2.31)–(2.33), (2.35)–(2.37), and Lemma 2.4 applied with $S = T_{r(n)}, u = x_n, \alpha = \alpha_n, M = 3M_0, p = p_\delta, \gamma = \epsilon_0$, and $v = x_{n+1}$, we have

$$d(x_{n+1}, p_\delta) \leq d(x_n, p_\delta) - \Delta_0. \tag{2.38}$$

Thus for each $n \in E_Q$ Eq. (2.38) holds.

Proposition 2.1, assumption (A), and (2.33) imply that for each $n \in \{0, \ldots, Q\}$,

$$d(x_{n+1}, p_\delta) = d((1 - \alpha_n)x_n \oplus \alpha_n T_{r(n)}(x_n), p_\delta)$$

$$\leq (1 - \alpha_n)d(x_n, p_\delta) + \alpha_n d(T_{r(n)}(x_n), p_\delta)$$

$$\leq (1 - \alpha_n)d(x_n, p_\delta) + \alpha_n d(x_n, p_\delta) + \delta = d(x_n, p_\delta) + \delta. \tag{2.39}$$

Assumption (A), (2.32), (2.35), (2.38) holding for each $n \in E_Q$ and (2.39) imply that

$$2M_0 \geq d(x_0, \theta) + d(\theta, p_\delta) \geq d(x_0, p_\delta) \geq d(x_0, p_\delta) - d(x_{Q+1}, p_\delta)$$

$$= \sum_{n=0}^{Q}(d(x_n, p_\delta) - d(x_{n+1}, p_\delta))$$

$$\geq \sum \{d(x_n, p_\delta) - d(x_{n+1}, p_\delta) : n \in E_Q\}$$

$$+ \sum \{d(x_n, p_\delta) - d(x_{n+1}, p_\delta) : \ n \in \{0, \ldots, Q\} \setminus E_Q\}$$

$$\geq \Delta_0 \mathrm{Card}(E_Q) - \delta(Q + 1),$$

$$\Delta_0 \mathrm{Card}(E_Q) \leq 2M_0 + \delta(Q + 1) \leq 2M_0 + 1,$$

$$\mathrm{Card}(E_Q) \leq \Delta_0^{-1}(2M_0 + 1). \tag{2.40}$$

Set

$$E = \{n \in \{0, 1, \ldots, \} : \ d(x_n, T_{r(n)}(x_n)) > \epsilon_0\}. \tag{2.41}$$

Since (2.40) holds for each natural number Q using (2.34) and (2.36), we conclude that

$$\mathrm{Card}(E) \leq \Delta_0^{-1}(2M_0 + 1)$$

$$= 4\bar{N}(2M_0 + 1)\epsilon^{-1}\Lambda^{-2}\eta^{-1}(6M_0 + 1, \epsilon\bar{N}^{-1}(6M_0 + 1)^{-1}). \tag{2.42}$$

Set

$$E_1 = \{n \in \{0, 1, \ldots\} : \ [n, n + \bar{N} - 1] \cap E \neq \emptyset\}. \tag{2.43}$$

By (2.42) and (2.43),

$$\mathrm{Card}(E_1) \leq \bar{N}\mathrm{Card}(E) \leq \Delta_0^{-1}(2M_0 + 1)\bar{N}$$

$$\leq 4\bar{N}^2(2M_0 + 1)\epsilon^{-1}\Lambda^{-2}\eta^{-1}(6M_0 + 1, \epsilon\bar{N}^{-1}(6M_0 + 1)^{-1}). \tag{2.44}$$

Assume that

$$n \in \{0, 1, \ldots\} \setminus E_1.$$

In view of (2.41) and (2.43),

$$[n, n + \bar{N} - 1] \cap E = \emptyset$$

and for each $i \in \{n, \ldots, n + \bar{N} - 1\}$,

$$d(x_i, T_{r(i)}(x_i)) \leq \epsilon_0. \tag{2.45}$$

Together with (2.33) and Proposition 2.1 this implies that

$$d(x_i, x_{i+1}) = d(x_i, (1 - \alpha_i)x_i \oplus \alpha_i T_{r(i)}(x_i)) \le \alpha_i d(x_i, T_{r(i)}(x_i)) \le \epsilon_0. \quad (2.46)$$

By (2.34) and (2.46), for each $i, j \in \{n, \ldots, n + \bar{N}\}$,

$$d(x_i, x_j) \le \bar{N}\epsilon_0 = \epsilon. \quad (2.47)$$

Let $s \in \{1, \ldots, m\}$. By the inclusion $r \in \mathcal{R}$, there exists $i \in \{n, \ldots, n + \bar{N} - 1\}$ such that $s = r(i)$. It follows from (2.34), (2.45), and (2.47) that

$$d(x_i, x_n) \le \epsilon,$$

$$d(x_i, T_s(x_i)) = d(x_i, T_{r(i)}(x_i)) \le \epsilon_0 = \epsilon \bar{N}^{-1}.$$

This implies that

$$x_n \in \tilde{F}_\epsilon.$$

Thus for each $n \in \{0, 1, \ldots\} \setminus E_1$, the inclusion above holds, and in view of (2.44),

$$\text{Card}(\{n \in \{0, 1, \ldots\} : x_n \notin \tilde{F}_\epsilon\}) \le \text{Card}(E_1)$$

$$\le 4\bar{N}^2(2M_0 + 1)\epsilon^{-1}\Lambda^{-2}\eta^{-1}(6M_0 + 1, \epsilon\bar{N}^{-1}(6M_0 + 1)^{-1}).$$

Theorem 2.8 is proved.

Theorem 2.9 *Assume that*

$$\Lambda \in (0, 2^{-1}), \quad (2.48)$$

$$d(T_i(x), T_i(y)) \le d(x, y), \ x, y \in C, \ i = 1, \ldots, m, \quad (2.49)$$

$\epsilon \in (0, 1), \ r \in \mathcal{R},$

$$r(i + \bar{N}) = r(i) \quad (2.50)$$

for each integer $i \ge 0$,

$$\{\alpha_i\}_{i=0}^\infty \subset [\Lambda, 1 - \Lambda], \quad (2.51)$$

$$\alpha_{i+\bar{N}} = \alpha_i \quad (2.52)$$

for each integer $i \geq 0$,

$$x_0 \in B(\theta, M_0) \cap C \tag{2.53}$$

and that for each integer $i \geq 0$,

$$x_{i+1} = (1 - \alpha_i)x_i \oplus \alpha_i T_{r(i)}(x_i). \tag{2.54}$$

Let

$$\epsilon_0 = 25^{-1}\bar{N}^{-1}(2\bar{N} + 1)^{-1}\epsilon \Lambda^2 \eta(6M_0 + 1, \epsilon(2\bar{N} + 1)^{-1}(6M_0 + 1)^{-1}), \tag{2.55}$$

$$N_0 = 4(2M_0 + 1)\bar{N}^2\epsilon_0^{-1}\Lambda^{-2}\eta^{-1}(6M_0 + 1, \epsilon_0\bar{N}^{-1}(6M_0 + 1)^{-1}). \tag{2.56}$$

Then for each integer $i \geq N_0\bar{N}$,

$$x_i \in F_\epsilon. \tag{2.57}$$

Proof Set

$$\epsilon_1 = \epsilon(2\bar{N} + 1)^{-1}. \tag{2.58}$$

In view of (2.56) and (2.58),

$$\epsilon_0 = 25^{-1}\bar{N}^{-1}\epsilon_1\Lambda^2\eta^{-1}(6M_0 + 1, \epsilon_1(6M_0 + 1)^{-1}). \tag{2.59}$$

For each integer $i \geq 0$ set

$$S_i(y) = (1 - \alpha_i)y \oplus \alpha_i T_{r(i)}(y), \quad y \in C. \tag{2.60}$$

By (2.47), (2.52), (2.54), and (2.60), for each integer $i \geq 0$,

$$S_{i+\bar{N}} = S_i, \quad x_{i+1} = S_i(x_i). \tag{2.61}$$

Theorem 2.8, Proposition 2.3, and Eqs. (2.51), (2.53), (2.54), and (2.56) imply that

$$\text{Card}(\{n \in \{0, 1, \ldots\} : x_n \notin F_{3\epsilon_0}\}) \leq N_0. \tag{2.62}$$

Set

$$E_0 = \{n \in \{0, 1, \ldots\} : x_n \notin F_{3\epsilon_0}\}, \tag{2.63}$$

$$E_1 = \{n \in \{0, 1, \ldots\} : [n, n + \bar{N} - 1] \cap E_0 \neq \emptyset\}. \tag{2.64}$$

By (2.62) and (2.64),

$$\mathrm{Card}(E_1) \leq \bar{N}\mathrm{Card}(E_0) \leq \bar{N}N_0. \tag{2.65}$$

In view of (2.65), there exists a nonnegative integer

$$n_0 \leq N_0\bar{N} \tag{2.66}$$

such that

$$\{n_0, \ldots, n_0 + \bar{N} - 1\} \cap E_0 = \emptyset. \tag{2.67}$$

Proposition 2.1 and (2.67) imply that for each $q \in \{n_0, \ldots, n_0 + \bar{N} - 1\}$ and each $j \in \{1, \ldots, m\}$,

$$d(x_q, T_j(x_q)) \leq 3\epsilon_0, \tag{2.68}$$

$$d(x_q, x_{q+1}) = d(x_q, (1 - \alpha_q)x_q \oplus \alpha_q T_{r(q)}(x_q)) \leq \alpha_q d(x_q, T_{r(q)}(x_q)) \leq 3\epsilon_0. \tag{2.69}$$

In view of (2.69), for each $i, j \in \{n_0, \ldots, n_0 + \bar{N}\}$,

$$d(x_i, x_j) \leq 3\bar{N}\epsilon_0$$

and in particular,

$$d(x_{n_0}, x_{n_0+\bar{N}}) \leq 3\bar{N}\epsilon_0. \tag{2.70}$$

Set

$$S = S_{n_0+\bar{N}-1} \cdots S_{n_0}. \tag{2.71}$$

Proposition 2.1 and Eqs. (2.46), (2.60), (2.62), and (2.71) imply that for each $x, y \in C$,

$$d(S(x), S(y)) \leq d(x, y), \tag{2.72}$$

$$x_{n_0+(i+1)\bar{N}} = S(x_{n_0+i\bar{N}}) \tag{2.73}$$

for each integer $i \geq 0$. It follows from (2.70), (2.71), and (2.73) that for each integer $i \geq 0$,

$$d(x_{n_0+(i+1)\bar{N}}, x_{n_0+i\bar{N}}) = d(S^i(x_{n_0+\bar{N}}), S^i(x_{n_0}))$$

$$\leq d(x_{n_0+\bar{N}}, x_{n_0}) \leq 3\bar{N}\epsilon_0. \tag{2.74}$$

Now we show that

$$d(x_i, T_{r(i)}(x_i)) \le \epsilon_1 \tag{2.75}$$

for each integer $i \ge n_0$.

Let $q \ge 0$ be an integer. In view of (2.74),

$$d(x_{n_0+q\bar{N}}, x_{n_0+(q+1)\bar{N}}) \le 3\bar{N}\epsilon_0. \tag{2.76}$$

Since q is any nonnegative integer in order to meet our goal, it is sufficient to show that (2.75) holds for each $i \in \{n_0 + q\bar{N}, \ldots, n_0 + (q+1)\bar{N} - 1\}$. Assume the contrary. Then there exists

$$j \in \{n_0 + q\bar{N}, \ldots, n_0 + (q+1)\bar{N} - 1\} \tag{2.77}$$

such that

$$d(x_j, T_{r(j)}(x_j)) > \epsilon_1. \tag{2.78}$$

Fix a positive number

$$\delta < \min\{(n_0+(q+1)\bar{N})^{-1}, \ 8^{-1}\epsilon_1 \Lambda^2 \eta(6M_0+1, \epsilon_1(6M_0+1)^{-1})\bar{N}^{-1}\}. \tag{2.79}$$

Proposition 2.1, assumption (A), and Eqs. (2.53) and (2.54) imply that for each integer $i \ge 0$,

$$d(x_{i+1}, p_\delta) = d((1 - \alpha_i)x_i \oplus \alpha_i T_{r(i)}(x_i), p_\delta)$$

$$\le (1 - \alpha_i)d(x_i, p_\delta) + \alpha_i d(T_{r(i)}(x_i), p_\delta)$$

$$\le (1 - \alpha_i)d(x_i, p_\delta) + \alpha_i d(x_i, p_\delta) + \delta = d(x_i, p_\delta) + \delta, \tag{2.80}$$

$$d(x_i, p_\delta) \le d(x_0, p_\delta) + i\delta \le 2M_0 + i\delta. \tag{2.81}$$

Proposition 2.5 implies that for each integer $n \ge 0$ and each $j \in \{1, \ldots, m\}$,

$$d(x_n, \theta) \le 3M_0, \ d(T^j(x_0), \theta) \le 3M_0. \tag{2.82}$$

By Eqs. (2.51), (2.54), (2.78), (2.79), (2.82) and Lemma 2.4 applied with $S = T_{r(j)}$, $u = x_j, \alpha = \alpha_j, M = 3M_0, p = p_\delta, \gamma = \epsilon_1$, and $v = x_{j+1}$, we have

$$d(x_{j+1}, p_\delta) \le d(x_j, p_\delta) - 4^{-1}\epsilon_1\alpha_j(1 - \alpha_j)\eta(6M_0+1, \epsilon_1(6M_0+1)^{-1}). \tag{2.83}$$

In view of (2.76), (2.79)–(2.81), and (2.83),

$$3\bar{N}\epsilon_0 \geq d(x_{n_0+q\bar{N}}, x_{n_0+(q+1)\bar{N}}) \geq d(p_\delta, x_{n_0+q\bar{N}}) - d(p_\delta, x_{n_0+(q+1)\bar{N}})$$

$$= \sum_{i=n_0+q\bar{N}}^{n_0+(q+1)\bar{N}-1} (d(x_i, p_\delta) - d(x_{i+1}, p_\delta))$$

$$\geq -\delta\bar{N} + d(x_j, p_\delta) - d(x_{j+1}, p_\delta)$$

$$\geq -\delta\bar{N} + 4^{-1}\epsilon_1 \Lambda^2 \eta(6M_0 + 1, \epsilon_1(6M_0 + 1)^{-1})$$

$$\geq 8^{-1}\epsilon_1 \Lambda^2 \eta(6M_0 + 1, \epsilon_1(6M_0 + 1)^{-1}).$$

This contradicts (2.59). The contradiction we have reached proves that

$$d(x_i, T_{r(i)}(x_i)) \leq \epsilon_1, \ i \in \{n_0 + q\bar{N}, \ldots, n_0 + (q+1)\bar{N} - 1\}.$$

Thus (2.75) holds for each integer $i \geq n_0$.

Proposition 2.1 and Eqs. (2.54) and (2.75) imply that for each integer $i \geq n_0$,

$$d(x_i, x_{i+1}) = d(x_i, (1-\alpha_i)x_i \oplus \alpha_i T_{r(i)}(x_i)) \leq \alpha_i d(x_i, T_{r(i)}(x_i)) \leq \epsilon_1$$

and for each $j \in \{i, \ldots, i + \bar{N}\}$,

$$d(x_i, x_j) \leq \epsilon_1 \bar{N}. \tag{2.84}$$

Assume that $i \geq n_0$ is an integer and $s \in \{1, \ldots, m\}$. By the inclusion $r \in \mathcal{R}$, there exists

$$j \in \{i, \ldots, i + \bar{N} - 1\} \text{ such that } s = r(j). \tag{2.85}$$

In view of (2.58), (2.75), (2.84), and (2.85),

$$d(x_j, T_s(x_j)) = d(x_j, T_{r(j)}(x_j)) \leq \epsilon_1,$$

$$d(x_i, T_s(x_i)) \leq d(x_i, x_j) + d(x_j, T_s(x_j)) + d(T_s(x_j), T_s(x_i))$$

$$\leq 2d(x_i, x_j) + \epsilon_1 \leq \epsilon_1(2\bar{N} + 1) \leq \epsilon,$$

$$x_i \in F_\epsilon$$

for each integer $i \geq n_0$. Theorem 2.9 is proved.

2.6   Inexact Iterates with Summable Errors

In this section we prove two theorems which describe the behavior of inexact iterates of our algorithm with summable errors.

Theorem 2.10 *Assume that* $\Lambda \in (0, 2^{-1})$, $\{\Delta_i\}_{i=0}^{\infty} \subset [0, \infty)$,

$$\Delta = \sum_{i=0}^{\infty} \Delta_i < \infty, \tag{2.86}$$

$$\{\alpha_i\}_{i=0}^{\infty} \subset (\Lambda, 1 - \Lambda), \tag{2.87}$$

$$x_0 \in B(\theta, M_0) \cap C, \ r \in \mathcal{R}, \tag{2.88}$$

for each integer $n \geq 0$,

$$x_{n+1} \in B((1 - \alpha_n)x_n \oplus \alpha_n T_{r(n)}(x_n), \Delta_n) \cap C \tag{2.89}$$

and $\epsilon \in (0, 1)$. *Let* n_0 *be a natural number such that for each integer* $i \geq n_0 \bar{N}$,

$$\Delta_i < \epsilon (2\bar{N})^{-1}. \tag{2.90}$$

Then

$$Card(\{n \in \{0, 1, \dots \} : \ there \ is \ i \in \{n, \dots, n + \bar{N}\}$$

$$such \ that \ x_i \notin \tilde{F}_\epsilon\})$$

$$\leq n_0\bar{N} + 8\bar{N}^2(4M_0 + 2\Delta)\epsilon^{-1}\eta^{-1}(6M_0 + 2\Delta + 1, \epsilon(2\bar{N})^{-1}(6M_0 + 2\Delta + 1)^{-1})\Lambda^{-2}.$$

Proof Set

$$\epsilon_0 = \epsilon \bar{N}^{-1}, \ \Delta_* = 4^{-1}\epsilon_0 \Lambda^2 \eta(6M_0 + 2\Delta + 1, \epsilon_0(6M_0 + 2\Delta + 1)^{-1}). \tag{2.91}$$

In view of Proposition 2.5, (2.88), and (2.89), for each integer $n \geq 0$, each $j \in \{1, \dots, m\}$, and each $\delta \in (0, 1)$,

$$d(x_n, \theta) \leq 3M_0 + \Delta, \ d(T_j(x_n), \theta) \leq 3M_0 + \Delta, \ d(x_{n+1}, p_\delta) \leq d(x_n, p_\delta) + \delta + \Delta_n. \tag{2.92}$$

By (2.90) and (2.91), for each integer $i \geq n_0 \bar{N}$,

$$\Delta_i < \epsilon_0. \tag{2.93}$$

Set

$$E = \{i \geq n_0 \bar{N} \text{ is an integer} : d(x_i, T_{r(i)}(x_i)) > \epsilon_0\}. \tag{2.94}$$

Let Q be a natural number. Choose a positive number

$$\delta \leq 4^{-1}\epsilon_0\eta(6M_0 + 1 + 2\Delta, \epsilon_0(6M_0 + 1 + 2\Delta)^{-1})\Lambda^2. \tag{2.95}$$

Assume that

$$j \in \{0, , \ldots, (n_0 + Q)\bar{N}\}, \tag{2.96}$$

$$d(x_j, T_{r(j)}(x_j)) > \epsilon_0. \tag{2.97}$$

By assumption (A), Eqs. (2.92), (2.97) and Lemma 2.4 applied with $S = T_{r(j)}$, $u = x_j$, $\alpha = \alpha_j$, $M = 3M_0 + \Delta$, $p = p_\delta$, $\gamma = \epsilon_0$, and

$$v = (1 - \alpha_j)x_j \oplus \alpha_j T_{r(j)}(x_j),$$

we have

$$d((1 - \alpha_j)x_j \oplus \alpha_j T_{r(j)}(x_j), p_\delta)$$

$$\leq d(x_j, p_\delta) - 4^{-1}\epsilon_0\Lambda^2\eta(6M_0 + 2\Delta + 1, \epsilon_0(6M_0 + 2\Delta + 1)^{-1}).$$

In view of the equation above, (2.89) and (2.91),

$$d(x_{j+1}, p_\delta) \leq d(x_{j+1}, (1 - \alpha_j)x_j \oplus \alpha_j T_{r(j)}(x_j))$$

$$+ d((1 - \alpha_j)x_j \oplus \alpha_j T_{r(j)}(x_j), p_\delta)$$

$$\leq \Delta_j + d(x_j, p_\delta) - 4^{-1}\epsilon_0\Lambda^2\eta(6M_0 + 2\Delta + 1, \epsilon_0(6M_0 + 2\Delta + 1)^{-1})$$

$$\leq \Delta_j + d(x_j, p_\delta) - \Delta_*.$$

Thus we have shown that the following property holds:
   (a) For each integer $j \in \{0, \ldots, (n_0 + Q)\bar{N}\}$ for which

$$d(x_j, T_{r(j)}(x_j)) > \epsilon_0$$

we have

$$d(x_{j+1}, p_\delta) \leq d(x_j, p_\delta) + \Delta_j - \Delta_*.$$

Set

$$E_Q = E \cap \{0, \dots, (n_0 + Q)\bar{N} - 1\}. \tag{2.98}$$

Property (a), assumption (A), and (2.92), (2.94), (2.98) imply that

$$4M_0 + \Delta \geq d(x_{n\bar{N}}, \theta) + d(\theta, p_\delta) \geq d(x_{n_0\bar{N}}, p_\delta) \geq d(x_{n_0\bar{N}}, p_\delta) - d(x_{(n_0+Q)\bar{N}}, p_\delta)$$

$$= \sum_{n=n_0\bar{N}}^{(n_0+Q)\bar{N}-1} (d(x_n, p_\delta) - d(x_{n+1}, p_\delta))$$

$$= \sum \{d(x_n, p_\delta) - d(x_{n+1}, p_\delta) : n \in E_Q\}$$

$$+ \sum \{d(x_n, p_\delta) - d(x_{n+1}, p_\delta) : n \in \{n_0\bar{N}, \dots, (n_0 + Q)\bar{N} - 1\} \setminus E_Q\}$$

$$\geq \sum \{\Delta_* - \Delta_n : n \in E_Q\} + \sum \{-\Delta_n - \delta : n \in \{n_0\bar{N}, \dots, (n_0+Q)\bar{N}-1\} \setminus E_Q\}$$

$$\geq \Delta_* \mathrm{Card}(E_Q) - \sum_{n_0\bar{N}}^{(n_0+Q)\bar{N}-1} -\delta Q\bar{N},$$

$$\mathrm{Card}(E_Q) \leq \Delta_*^{-1}(4M_0 + 2\Delta + \delta Q\bar{N}).$$

Since δ is any positive number satisfying (2.95), we conclude that

$$\mathrm{Card}(E_Q) \leq \Delta_*^{-1}(4M_0 + 2\Delta).$$

Since Q is any natural number, we conclude that

$$\mathrm{Card}(E) \leq \Delta_*^{-1}(4M_0 + 2\Delta). \tag{2.99}$$

Set

$$E_0 = \{n \geq n_0\bar{N} \text{ is an integer and } [n, n + \bar{N} - 1] \cap E \neq \emptyset\}. \tag{2.100}$$

By (2.91), (2.99), and (2.100),

$$\mathrm{Card}(E_0) \leq \bar{N}\mathrm{Card}(E) \leq \Delta_*^{-1}(4M_0 + 2\Delta)\bar{N}$$

$$\leq 8\bar{N}^2(4M_0 + 2\Delta)\epsilon^{-1}\Lambda^{-2}\eta^{-1}(6M_0 + 2\Delta + 1, \epsilon(2\bar{N}^{-1})(6M_0 + 2\Delta + 1)^{-1}). \tag{2.101}$$

Assume that

$$n \in \{n_0\bar{N}, n_0\bar{N} + 1, \ldots\} \setminus E_0. \tag{2.102}$$

In view of (2.100) and (2.102),

$$[n, n + \bar{N} - 1] \cap E = \emptyset$$

and for each $i \in \{n, \ldots, n + \bar{N} - 1\}$,

$$d(x_i, T_{r(i)}(x_i)) \leq \epsilon_0. \tag{2.103}$$

Let $i \in \{n, \ldots, n + \bar{N} - 1\}$. By (2.89), (2.93), (2.102), (2.103) and Proposition 2.1,

$$d(x_i, x_{i+1}) \leq d(x_i, (1 - \alpha_i)x_i \oplus \alpha_i T_{r(i)}(x_i))$$

$$+ d((1 - \alpha_i)x_i \oplus \alpha_i T_{r(i)}(x_i), x_{i+1})$$

$$\leq \alpha_i d(x_i, T_{r(i)}(x_i)) + \Delta_i \leq \epsilon_0 + \Delta_i \leq 2\epsilon_0. \tag{2.104}$$

By (2.104), for each $i, j \in \{n, \ldots, n + \bar{N}\}$,

$$d(x_i, x_j) \leq 2\bar{N}\epsilon_0. \tag{2.105}$$

Let $i \in \{n, \ldots, n + \bar{N}\}$ and $s \in \{1, \ldots, m\}$. By (2.88), there exists $j \in \{n, \ldots, n + \bar{N} - 1\}$ such that $s = r(j)$. It follows from (2.103) that

$$d(x_j, T_s(x_j)) = d(x_j, T_{r(j)}(x_j)) \leq \epsilon_0.$$

Together with (2.91), (2.103), and (2.105), this implies that

$$x_i \in \tilde{F}_\epsilon, \quad i \in \{n, \ldots, n + \bar{N}\}.$$

Thus for each integer $n \geq n_0\bar{N}$ satisfying $n \notin E_0$ and each $i \in \{n, \ldots, n + \bar{N}\}$, we have

$$x_i \in \tilde{F}_\epsilon.$$

Then

$$\tilde{E} := \{n \in \{0, 1, \ldots\} : \quad \text{there is } i \in \{n, \ldots, n + \bar{N}\} \text{ for which } x_i \notin \tilde{F}_\epsilon\}$$

$$\subset \{0, \ldots, n_0\bar{N} - 1\} \cup E_0,$$

and by (2.101),

$$\mathrm{Card}(\tilde{E}) \leq n_0\bar{N} + \mathrm{Crad}(E) \leq n_0\bar{N}$$

$$+ 8\bar{N}^2(4M_0 + 2\Delta)\epsilon^{-1}\Lambda^{-2}\eta^{-1}(6M_0 + 2\Delta + 1, \epsilon(2\bar{N}^{-1})(6M_0 + 2\Delta + 1)^{-1}).$$

Theorem 2.10 is proved.

Theorem 2.11 *Assume that*

$$\Lambda \in (0, 2^{-1}),$$

$$\{\Delta_i\}_{i=0}^\infty \subset [0, \infty),$$

$$\Delta = \sum_{i=0}^\infty \Delta_i < \infty, \tag{2.106}$$

$$d(T_i(x), T_i(y)) \leq d(x, y), \ x, y \in C, \ i = 1, \dots, m, \tag{2.107}$$

$\epsilon \in (0, 1), r \in \mathcal{R},$

$$r(i + \bar{N}) = r(i) \tag{2.108}$$

for each integer $i \geq 0,$

$$\{\alpha_i\}_{i=0}^\infty \subset [\Lambda, 1 - \Lambda], \tag{2.109}$$

$$\alpha_{i+\bar{N}} = \alpha_i \tag{2.110}$$

for each integer $i \geq 0,$

$$x_0 \in B(\theta, M_0) \cap C, \tag{2.111}$$

for each integer $i \geq 0,$

$$x_{i+1} \in B((1 - \alpha_i)x_i \oplus \alpha_i T_{r(i)}(x_i), \Delta_i), \tag{2.112}$$

n_0 is a natural number and that

$$\sum_{i=n_0}^\infty \Delta_i \leq \epsilon/4. \tag{2.113}$$

Let

$$\epsilon_0 = 10^{-2}\bar{N}^{-1}(2\bar{N}+1)^{-1}\epsilon\Lambda^2$$

$$\times \eta(18M_0 + 6\Delta + 1, 4^{-1}\epsilon(2\bar{N}+1)^{-1}(18M_0 + 6\Delta + 1)^{-1}), \tag{2.114}$$

$$N_0 = 4(6M_0 + 2\Delta + 1)\bar{N}^2\epsilon_0^{-2}\Lambda^{-2}\eta^{-1}(18M_0 + 6\Delta + 1, \epsilon_0\bar{N}^{-1}(18M_0 + 6\Delta + 1)^{-1}). \tag{2.115}$$

Then for each integer $i \geq n_0 + N_0\bar{N}$,

$$x_i \in F_\epsilon.$$

Proof Proposition 2.5 and Eqs. (2.106), (2.108), (2.111), and (2.112) imply that for each integer ≥ 0,

$$\rho(x_i, \theta) \leq 3M_0 + \Delta. \tag{2.116}$$

Define a sequence $\{y_i\}_{i=n_0}^\infty \subset X$ such that

$$y_{n_0} = x_{n_0} \tag{2.117}$$

and that for each integer $i \geq n_0$,

$$y_{i+1} = (1 - \alpha_i)y_i \oplus \alpha_i T_{r(i)}(y_i). \tag{2.118}$$

Theorem 2.9 and Eqs. (2.107)–(2.110), (2.116)–(2.118) imply that for each integer $i \geq n_0 + N_0\bar{N}$,

$$y_i \in F_{\epsilon/4}. \tag{2.119}$$

We show that for each integer $i \geq n_0 + 1$,

$$d(x_i, y_i) \leq \sum_{j=n_0}^{i-1} \Delta_j. \tag{2.120}$$

By (2.112), (2.117), and (2.118),

$$d(x_{n_0+1}, y_{n_0+1}) = d(x_{n_0+1}, (1 - \alpha_{n_0})x_{n_0} \oplus \alpha_{n_0}T_{r(n_0)}(x_{n_0})) \leq \Delta_{n_0},$$

and (2.120) holds for $i = n_0 + 1$.

Assume that $k \geq n_0 + 1$ is an integer and (2.120) holds for $i = k$. Proposition 2.1, (2.107), (2.112), (2.118), and (2.120) imply that

$$d(x_{k+1}, y_{k+1}) \leq d(x_{k+1}, (1 - \alpha_k)x_k \oplus \alpha_k T_{r(k)}(x_k))$$

$$+ d((1 - \alpha_k)x_k \oplus \alpha_k T_{r(k)}(x_k), (1 - \alpha_k)y_k \oplus \alpha_k T_{r(k)}(y_k))$$

$$\leq \Delta_k + (1 - \alpha_k)d(x_k, y_k) + \alpha_k d(T_{r(k)}(x_k), T_{r(k)}(y_k))$$

$$\leq \Delta_k + d(x_k, y_k) \leq \sum_{j=n_0}^{k} \Delta_j,$$

and (2.120) holds for $k + 1$ too. Thus by induction we showed that (2.120) holds for each integer $i \geq n_0 + 1$. Together with (2.103) this implies that for all integers $i \geq n_0$,

$$d(x_i, y_i) \leq \epsilon/4. \tag{2.121}$$

Assume that $i \geq n_0 + N_0 \bar{N}$ is an integer and $s \in \{1, \ldots, m\}$. In view of (2.119),

$$d(y_i, T_s(y_i)) \leq \epsilon/4.$$

Together with (2.107) and (2.121), this implies that

$$d(x_i, T_s(x_i)) \leq d(x_i, y_i) + d(y_i, T_s(y_i)) + d(T_s(y_i), T_s(x_i))$$

$$\leq 2d(x_i, y_i) + d(y_i, T_s(y_i)) \leq \epsilon/2 + \epsilon/4 < \epsilon$$

and $x_i \in F_\epsilon$. Theorem 2.11 is proved.

2.7   Inexact Iterates with Nonsummable Errors

In this section we prove two theorems which describe the behavior of inexact iterates of our algorithm with nonsummable errors. These results show that if computational errors are small enough, then our method generates approximate solution belonging to the set $\tilde{F}_\gamma$ with $\gamma > 0$. Our first result shows the dependence of γ on our computational errors and calculates the number of iterates which should be done in order to obtain this approximate solution.

Theorem 2.12 *Let* $\Lambda \in (0, 2^{-1})$, $\epsilon_0 \in (0, 1/4)$,

$$0 < \delta_0 \leq 32^{-1} \bar{N}^{-1} \epsilon_0 \Lambda^2 \eta(6M_0 + 3, \epsilon_0(6M_0 + 3)^{-1}), \tag{2.122}$$

$$n_0 = \lfloor 32M_0 \epsilon_0^{-1} \Lambda^{-2} \eta^{-1}(6M_0 + 3, \epsilon_0(6M_0 + 3)^{-1}) \rfloor + 1. \tag{2.123}$$

Assume that

$$r \in \mathcal{R}, \ \{\alpha_i\}_{i=0}^{\infty} \subset [\Lambda, 1 - \Lambda], \tag{2.124}$$

$$x_0 \in B(\theta, M_0) \cap C, \tag{2.125}$$

and that for each integer $n \geq 0$,

$$x_{n+1} \in B((1 - \alpha_n)x_n \oplus \alpha_n T_{r(n)}(x_n), \delta_0). \tag{2.126}$$

Then there exists an integer $q \in [0, n_0 - 1]$ such that

$$d(x_i, \theta) \leq 3M_0 + 1, \ i = 0, \ldots, q\bar{N},$$

$$d(x_i, T_{r(i)}(x_i)) \leq \epsilon_0, \ i = q\bar{N}, \ldots, (q+1)\bar{N} - 1. \tag{2.127}$$

Moreover, if $q \in \{0, \ldots, n_0 - 1\}$ and (2.127) holds, then for each $i \in \{q\bar{N}, \ldots, (q+1)\bar{N} - 1\}$,

$$d(x_i, x_{i+1}) \leq \epsilon_0 + \delta_0$$

and

$$x_i \in \tilde{F}_{(\epsilon_0 + \delta_0)\bar{N}}, \ i \in \{q\bar{N}, \ldots, (q+1)\bar{N}\}.$$

Proof Assume that s is a nonnegative integer and that for each $k \in [0, s]$,

$$\max\{d(x_{i-1}, T_{r(i-1)}(x_{i-1})) : \ i = k\bar{N} + 1, \ldots, (k+1)\bar{N}\} > \epsilon_0. \tag{2.128}$$

Fix

$$\delta \in (0, 2^{-1}\delta_0). \tag{2.129}$$

Assumption (A) and (2.125) imply that

$$d(x_0, p_\delta) \leq 2M. \tag{2.130}$$

Assume that $k \in [0, s]$ as an integer and

$$d(x_{k\bar{N}}, p_\delta) \leq 2M_0. \tag{2.131}$$

Lemma 2.13 *Assume that $i \in \{0, \ldots, \bar{N} - 1\}$ and that*

$$d(x_{k\bar{N}+i}, p_\delta) \leq 2M_0 + 2i\delta_0. \tag{2.132}$$

Then

$$d(x_{k\bar{N}+i+1}, p_\delta) \leq d(x_{k\bar{N}+i}, p_\delta) + 2\delta_0.$$

If $d(x_{k\bar{N}+i}, T_{r(k\bar{N}+i)}(x_{k\bar{N}+i})) > \epsilon_0$, then

$$d(x_{k\bar{N}+i+1}, p_\delta) - d(x_{k\bar{N}+i}, p_\delta) < -8^{-1}\epsilon_0\Lambda^2\eta(6M_0 + 3, \epsilon_0(6M_0 + 3)^{-1}).$$

Proof Proposition 2.1, assumption (A), and Eqs. (2.126), (2.129) imply that

$$d(x_{k\bar{N}+i+1}, p_\delta) \leq d(x_{k\bar{N}+i+1}, (1 - \alpha_{k\bar{N}+i})x_{k\bar{N}+i} \oplus \alpha_{k\bar{N}+i}T_{r(k\bar{N}+i)}(x_{k\bar{N}+i}))$$

$$+ d((1 - \alpha_{k\bar{N}+i})x_{k\bar{N}+i} \oplus \alpha_{k\bar{N}+i}T_{r(k\bar{N}+i)}(x_{k\bar{N}+i}), p_\delta)$$

$$\leq \delta_0 + (1 - \alpha_{k\bar{N}+i})d(x_{k\bar{N}+i}, p_\delta) + \alpha_{k\bar{N}+i}d(T_{r(k\bar{N}+i)}(x_{k\bar{N}+i}), p_\delta)$$

$$\leq \delta_0 + d(x_{k\bar{N}+i}, p_\delta) + \delta \leq d(x_{k\bar{N}+i}, p_\delta) + 2\delta_0. \tag{2.133}$$

Assume that

$$d(x_{k\bar{N}+i}, T_{r(k\bar{N}+i)}(x_{k\bar{N}+i})) > \epsilon_0. \tag{2.134}$$

Assumption (A), (2.122), and (2.132) imply that

$$d(x_{k\bar{N}+i}, \theta) \leq d(x_{k\bar{N}+i}, p_\delta) + d(p_\delta, \theta) \leq 2M_0 + 2\bar{N}\delta_0 + M_0 \leq 3M_0 + 1. \tag{2.135}$$

By assumption (A), Eqs. (2.122), (2.124), (2.129), (2.134), (2.135), and Lemma 2.4 applied with $S = T_{r(k\bar{N}+i)}, u = x_{k\bar{N}+i}, \alpha = \alpha_{k\bar{N}+i}, M = 3M_0+1, p = p_\delta, \gamma = \epsilon_0$, and

$$v = (1 - \alpha_{k\bar{N}+i})x_{k\bar{N}+i} \oplus \alpha_{k\bar{N}+i}T_{r(k\bar{N}+i)}(x_{k\bar{N}+i}),$$

we have

$$d((1 - \alpha_{k\bar{N}+i})x_{k\bar{N}+i} \oplus \alpha_{k\bar{N}+i}T_{r(k\bar{N}+i)}(x_{k\bar{N}+i}), p_\delta)$$

$$\leq d(x_{k\bar{N}+i}, p_\delta) - 4^{-1}\epsilon_0\Lambda^2\eta(6M_0 + 3, \epsilon_0(6M_0 + 3)^{-1}).$$

Together with (2.122) and (2.126), this implies that

$$d(x_{k\bar{N}+i+1}, p_\delta) \leq d((1 - \alpha_{k\bar{N}+i})x_{k\bar{N}+i} \oplus \alpha_{k\bar{N}+i}T_{r(k\bar{N}+i)}(x_{k\bar{N}+i}), p_\delta)$$

$$+ d(x_{k\bar{N}+i+1}, (1 - \alpha_{k\bar{N}+i})x_{k\bar{N}+i} \oplus \alpha_{k\bar{N}+i}T_{r(k\bar{N}+i)}(x_{k\bar{N}+i}))$$

$$\leq d(x_{k\bar{N}+i}, p_\delta) - 4^{-1}\epsilon_0\Lambda^2\eta(6M_0 + 3, \epsilon_0(6M_0 + 3)^{-1}) + \delta_0$$

$$\leq d(x_{k\bar{N}+i}, p_\delta) - 8^{-1}\epsilon_0\Lambda^2\eta(6M_0 + 3, \epsilon_0(6M_0 + 3)^{-1}).$$

Lemma 2.13 is proved.

It follows from Eqs. (2.131) and (2.122) and Lemma 2.13 applied by induction that for all $i = 0, \ldots, \bar{N} - 1$,

$$d(x_{k\bar{N}+i+1}, p_\delta) \le d(x_{k\bar{N}+i}, p_\delta) + 2\delta_0, \qquad (2.136)$$

$$d(x_{k\bar{N}+i+1}, p_\delta) \le 2M + 2\delta_0(i + 1) \le 2M + 2\delta_0\bar{N} \le 2M + 1,$$

$$d(x_{k\bar{N}+i}, p_\delta) \le 2M + 1, \ i = 0, \ldots, \bar{N}.$$

By (2.122), (2.128), and (2.136),

$$d(x_{(k+1)\bar{N}}, p_\delta) - d(x_{k\bar{N}}, p_\delta) = \sum_{i=0}^{\bar{N}-1} (d(x_{k\bar{N}+i+1}, p_\delta) - d(x_{k\bar{N}+i}, p_\delta))$$

$$\le -8^{-1}\epsilon_0\Lambda^2\eta(6M_0 + 3, \epsilon_0(6M_0 + 3)^{-1}) + 2\delta_0\bar{N}$$

$$\le -16^{-1}\epsilon_0\Lambda^2\eta(6M_0 + 3, \epsilon_0(6M_0 + 3)^{-1}).$$

Thus we have shown that the following property holds:

(P1)   If an integer $k \in [0, s]$ satisfies $d(x_{k\bar{N}}, p_\delta) \le 2M_0$, then

$$d(x_{k\bar{N}+i}, p_\delta) \le 2M + 1, \ i = 0, \ldots, \bar{N}, \qquad (2.137)$$

$$d(x_{(k+1)\bar{N}}, p_\delta) \le d(x_{k\bar{N}}, p_\delta) - 16^{-1}\epsilon_0\Lambda^2\eta(6M_0 + 3, \epsilon_0(6M_0 + 3)^{-1}). \qquad (2.138)$$

By (2.130) and property (P1),

$$d(x_j, p_\delta) \le 2M_0 + 1, \ j = 0, \ldots, (s + 1)\bar{N}$$

and (2.138) holds for $k = 0, \ldots, s$. In view of (2.130) and (2.138),

$$16^{-1}\epsilon_0\Lambda^2\eta(6M_0 + 3, \epsilon_0(6M_0 + 3)^{-1})(s + 1) \le \sum_{k=0}^{s}(d(x_{k\bar{N}}, p_\delta) - d(x_{(k+1)\bar{N}}, p_\delta))$$

$$\le d(x_0, p_\delta) - d(x_{(s+1)\bar{N}}, p_\delta) \le d(x_0, p_\delta) \le 2M_0,$$

$$s + 1 \le 32M_0\epsilon_0^{-1}\Lambda^{-2}\eta^{-1}(6M_0 + 3, \epsilon_0(6M_0 + 3)^{-1}).$$

Thus we have shown that the following property holds:

(P2)   If an integer $s \geq 0$ and for each $k \in \{0, \ldots, s\}$ (2.128) holds, then

$$s \leq 32 M_0 \epsilon_0^{-1} \Lambda^{-2} \eta^{-1} (6M_0 + 3, \epsilon_0 (6M_0 + 3)^{-1}) - 1,$$

$$d(x_j, p_\delta) \leq 2M_0 + 1, \ j = 0, \ldots, (s+1)\bar{N},$$

$$d(x_{k\bar{N}}, p_\delta) \leq 2M_0, \ k = 0, \ldots, (s+1).$$

Property (P2) and (2.123) imply that there exists an integer $q \in [0, n_0 - 1]$ such that for each integer k satisfying $0 \leq k < q$,

$$\max\{d(x_{i-1}, T_{r(i-1)}(x_{i-1})) : \ i = k\bar{N} + 1, \ldots, (k+1)\bar{N}\} > \epsilon_0,$$

$$\max\{d(x_{i-1}, T_{r(i-1)}(x_{i-1})) : \ i = q\bar{N} + 1, \ldots, (q+1)\bar{N}\} \leq \epsilon_0,$$

$$d(x_{q\bar{N}}, p_\delta) \leq 2M_0, \ d(x_j, p_\delta) \leq 2M_0 + 1, \ j = 0, \ldots, q\bar{N},$$

$$d(x_j, \theta) \leq 3M_0 + 1, \ j = 0, \ldots, q\bar{N}.$$

Assume that $q \in \{0, \ldots, n_0 - 1\}$ satisfies

$$d(x_{i-1}, T_{r(i-1)}(x_{i-1})) \leq \epsilon_0, \ i = q\bar{N} + 1, \ldots, (q+1)\bar{N}. \tag{2.139}$$

Proposition 2.1 and Eqs. (2.126) and (2.139) imply that for $i \in \{q\bar{N} + 1, \ldots, (q+1)\bar{N}\}$,

$$d(x_i, x_{i-1}) \leq d(x_i, (1 - \alpha_{i-1})x_{i-1} \oplus \alpha_{i-1} T_{r(i-1)}(x_{i-1}))$$

$$+ d((1 - \alpha_{i-1})x_{i-1} \oplus \alpha_{i-1} T_{r(i-1)}(x_{i-1}), x_{i-1})$$

$$\leq \delta_0 + \alpha_{i-1} d(x_{i-1}, T_{r(i-1)}(x_{i-1})) \leq \epsilon_0 + \delta_0.$$

This implies that for each $i, j \in \{q\bar{N}, \ldots, (q+1)\bar{N}\}$,

$$d(x_i, x_j) \leq (\epsilon_0 + \delta_0)\bar{N}. \tag{2.140}$$

Assume that $i \in \{q\bar{N}, \ldots, (q+1)\bar{N}\}$ and $s \in \{1, \ldots, m\}$. By (2.124), there exists $j \in \{q\bar{N}, \ldots, (q+1)\bar{N} - 1\}$ such that $s = r(j)$. By (2.139),

$$d(x_j, T_s(x_j)) \leq \epsilon_0.$$

Together with (2.140) this implies that $x_i \in \tilde{F}_{\bar{N}(\epsilon_0 + \delta_0)}$ and completes the proof of Theorem 2.12.

Theorem 2.12 implies the following result.

Theorem 2.14 *Assume that $\bar{\epsilon} \in (0, 1)$, $\tilde{F}_{\bar{\epsilon}} \subset B(\theta, M_0)$,*

$$0 < \epsilon_0 < (2\bar{N})^{-1}\bar{\epsilon},$$

$$\Lambda \in (0, 2^{-1}),$$

$$0 < \delta_0 \leq 32^{-1}\bar{N}^{-1}\epsilon_0 \Lambda^2 \eta(6M_0 + 3, \epsilon_0(6M_0 + 3)^{-1}),$$

$$n_0 = \lfloor 32M_0\epsilon_0^{-1}\Lambda^{-2}\eta^{-1}(6M_0 + 3, \epsilon_0(6M_0 + 3)^{-1})\rfloor + 1.$$

Assume that

$$r \in \mathcal{R}, \ \{\alpha_i\}_{i=0}^{\infty} \subset (\Lambda, 1 - \Lambda),$$

$$x_0 \in B(\theta, M_0) \cap C,$$

for each integer $n \geq 0$,

$$x_{n+1} \in B((1 - \alpha_n)x_n \oplus \alpha_n T_{r(n)}(x_n), \delta_0).$$

Then

$$d(x_i, \theta) \leq 3M_0 + 1$$

for all integers $i \geq 0$, and there exists a strictly increasing sequence of integers $\{q_p\}_{p=0}^{\infty}$ such that

$$0 \leq q_0 \leq n_0,$$

$$1 \leq q_{p+1} - q_p \leq n_0$$

for each integer $p \geq 0$ and that for each integer $p \geq 0$ and each $i \in \{q_p\bar{N}, \ldots, (q_p + 1)\bar{N}\}$,

$$x_i \in \tilde{F}_{2\epsilon_0\bar{N}}.$$

Theorem 2.15 *Assume that $\bar{\epsilon} \in (0, 1)$,*

$$F_{\bar{\epsilon}} \subset B(\theta, M_0), \tag{2.141}$$

$$\Lambda \in (0, 2^{-1}),$$

$$d(T_i(x), T_i(y)) \leq d(x, y), \ x, y \in C, \ i = 1, \ldots, m, \tag{2.142}$$

$$\epsilon \in (0, \bar{\epsilon}), \ \epsilon_1 = \epsilon/2,$$

$$\epsilon_0 = 25^{-1}\bar{N}^{-1}(2\bar{N}+1)^{-1}\epsilon_1\Lambda^2\eta(6M_0+1, \epsilon_1(2\bar{N}+1)^{-1}(6M_0+1)^{-1}), \quad (2.143)$$

$$N_0 = \lfloor 4(2M_0+1)\bar{N}^2\epsilon_0^{-1}\Lambda^{-2}\eta^{-1}(6M_0+1, \epsilon_0\bar{N}^{-1}(6M_0+1)^{-1})\rfloor +1, \quad (2.144)$$

$$\delta = (8N_0\bar{N})^{-1}\epsilon, \quad (2.145)$$

$r \in \mathcal{R}$,

$$r(i + \bar{N}) = r(i) \quad (2.146)$$

for each integer $i \geq 0$,

$$\{\alpha_i\}_{i=0}^{\infty} \subset [\Lambda, 1 - \Lambda], \quad (2.147)$$

$$\alpha_{i+\bar{N}} = \alpha_i \quad (2.148)$$

for each integer $i \geq 0$,

$$x_0 \in B(\theta, M_0) \cap C \quad (2.149)$$

and that for each integer $n \geq 0$,

$$x_{n+1} \in B((1 - \alpha_n)x_n \oplus \alpha_n T_{r(n)}(x_n), \delta). \quad (2.150)$$

Then for each integer $i \geq N_0\bar{N}$,

$$x_i \in F_\epsilon.$$

Proof Set

$$y_0 = x_0, \quad (2.151)$$

and for each integer $i \geq 0$,

$$y_{i+1} = (1 - \alpha_i)y_i \oplus \alpha_i T_{r(i)}(y_i). \quad (2.152)$$

Proposition 2.5 and Eqs. (2.149), (2.151), and (2.152) imply that for each integer $i \geq 0$,

$$\rho(y_i, \theta) \leq 3M_0, \quad d(T^j(y_i), \theta) \leq 3M_0, \quad j = 1, \ldots, m. \quad (2.153)$$

Theorem 2.9 and Eqs. (2.142), (2.143), (2.146)–(2.149), (2.151) and (2.152) imply that for each integer $i \geq N_0\bar{N}$,

$$y_i \in F_{\epsilon_1}. \quad (2.154)$$

We show that for each integer $i \geq 0$,

$$d(x_i, y_i) \leq i\delta. \tag{2.155}$$

Clearly, for $i = 0$, (2.155) holds.

Assume that $k \geq 0$ is an integer and (2.155) holds for $i = k$. Proposition 2.1, (2.142), (2.150), (2.152), and (2.155) imply that

$$d(x_{k+1}, y_{k+1}) \leq d(x_{k+1}, (1 - \alpha_k)x_k \oplus \alpha_k T_{r(k)}(x_k))$$

$$+ d((1 - \alpha_k)x_k \oplus \alpha_k T_{r(k)}(x_k), (1 - \alpha_k)y_k \oplus \alpha_k T_{r(k)}(y_k))$$

$$\leq (1 - \alpha_k)d(x_k, y_k) + \alpha_k d(T_{r(k)}(y_k), T_{r(k)}(x_k)) + \delta$$

$$\leq \delta + d(x_k, y_k) \leq \delta(k + 1)$$

and (2.155) holds for $k + 1$ too. Thus by induction we showed that (2.155) holds for each integer $i \geq 0$.

Let

$$i \in \{N_0\bar{N}, \ldots, 2N_0\bar{N}\}. \tag{2.156}$$

In view of (2.155) and (2.156),

$$d(x_i, y_i) \leq 2N_0\bar{N}\delta. \tag{2.157}$$

By (2.142), (2.145), (2.154)–(2.156), for each $s \in \{1, \ldots, m\}$,

$$d(x_i, T_s(x_i)) \leq d(x_i, y_i) + d(y_s, T_s(y_i)) + d(T_s(y_i), T_s(x_i))$$

$$\leq 2d(x_i, y_i) + \epsilon \leq 4N_0\bar{N}\delta + \epsilon_1 \leq \epsilon.$$

Thus

$$x_i \in F_\epsilon, \quad i \in \{N_0\bar{N}, \ldots, 2N_0\bar{N}\}. \tag{2.158}$$

Together with (2.141) this implies that

$$x_i \in B(\theta, M_0), \quad i \in \{N_0\bar{N}, \ldots, 2N_0\bar{N}\}.$$

Assume that $q \geq 0$ is an integer and that

$$x_i \in F_\epsilon, \quad i \in \{q, \ldots, q + N_0\bar{N}\}. \tag{2.159}$$

(In view of (2.158), relation (2.159) holds for $q = N_0\bar{N}$.) Set

$$y_q = x_q, \tag{2.160}$$

and for each integer $i \geq q$,

$$y_{i+1} = (1 - \alpha_i)y_i \oplus \alpha_i T_{r(i)}(y_i). \tag{2.161}$$

We show that for each integer $i \geq 0$,

$$d(x_{q+i}, y_{q+i}) \leq i\delta. \tag{2.162}$$

In view of (2.160), Eq. (2.162) holds for $i = 0$.

Assume that $k \geq 0$ is an integer and (2.162) holds for $i = k$. Proposition 2.1, (2.142), (2.150), (2.161), and (2.162) imply that

$$d(x_{q+k+1}, y_{q+k+1}) \leq d(x_{q+k+1}, (1 - \alpha_{q+k})x_{q+k} \oplus \alpha_{q+k}T_{r(q+k)}(x_{q+k}))$$

$$+ d((1 - \alpha_{q+k})x_{q+k} \oplus \alpha_{q+k}T_{r(q+k)}(x_{q+k}), (1 - \alpha_{q+k})y_{q+k} \oplus \alpha_{q+k}T_{r(q+k)}(y_{q+k}))$$

$$\leq (1 - \alpha_{q+k})d(x_{q+k}, y_{q+k}) + \alpha_{q+k}d(T_{r(q+k)}(y_{q+k}), T_{r(q+k)}(x_{q+k})) + \delta$$

$$\leq \delta + d(x_k, y_k) \leq \delta(k + 1),$$

and (2.162) holds for $k + 1$ too. Thus by induction we showed that (2.162) holds for each integer $i \geq 0$. Theorem 2.9 applied to the sequence $\{y_{q+i}\}_{i=0}^{\infty}$ and Eqs. (2.141), (2.142), (2.143), (2.146), (2.147) and (2.159)–(2.161) imply that for each integer $i \geq N_0\bar{N}$,

$$y_{i+q} \in F_{\epsilon_1}. \tag{2.163}$$

Assume that

$$i \in \{N_0\bar{N}, \ldots, 2N_0\bar{N}\}, \quad s \in \{1, \ldots, m\}. \tag{2.164}$$

By (2.142), (2.144), (2.162)–(2.164),

$$d(x_{q+i}, T_s(x_{q+i})) \leq d(x_{q+i}, y_{q+i}) + d(y_{q+i}, T_s(y_{q+i}))$$

$$+ d(T_s(y_{q+i}), T_s(x_{q+i})) \leq 2(x_{q+i}, y_{q+i}) + \epsilon_1 \leq 4N_0\bar{N}\delta + \epsilon_1 \leq \epsilon.$$

Thus

$$x_i \in F_\epsilon, \quad i \in \{q + N_0\bar{N}, \ldots, q + 2N_0\bar{N}\}.$$

This completes the proof of Theorem 2.14.

Chapter 3
Methods with Remotest Set Control

Abstract In this chapter we study the convergence of methods with remotest set control for solving common fixed point problems in a W-hyperbolic space using Krasnoselskii-Mann iterations. Our main goal is to obtain an approximate solution of the problem in the presence of computational errors. We show that the iterative method generates a good approximate solution, if the sequence of computational errors is bounded from above by a constant. Moreover, for a known computational error, we find out what an approximate solution can be obtained and how many iterates one needs for this.

3.1 Inexact Iterates

Assume that $X = (X, d, W)$ is a W-hyperbolic space with a structure (X, η), where $\eta : (0, \infty) \times (0, 2] \to (0, 1]$ is its modulus of uniform convexity. We continue to use the notation and definitions introduced in Sects. 2.1–2.3. In particular, we assume that $C \subset X$ is a non-empty convex set, $\theta \in C$, m is a natural number, and that for $i = 1, \ldots, m$, $T_i : C \to C$.

Assume that $M_0 > 0$ and the following assumption holds:

(A) For each $\delta \in (0, 1)$ there exists $p_\delta \in B(\theta, M_0) \cap C$ such that for each $i \in \{1, \ldots, m\}$ and each $x \in C$,

$$d(T_i(x), p_\delta) \le d(x, p_\delta) + \delta.$$

For each $\delta \in (0, 1)$ let p_δ be as guaranteed by (A).

Now we describe our algorithm.

Initialization: Choose

$$x \in C \cap B(\theta, M_0)$$

and set $x_0 = x$.

Iterative step: For each integer $i \geq 0$ choose $\alpha_i \in (0, 1)$ and $r(i) \in \{1, \ldots, m\}$ such that

$$d(x_i, T_{r(i)}(x_i)) = \max\{d(x_i, T_s(x_i)) : s = 1, \ldots, m\}$$

and set $x_{i+1} = (1 - \alpha_i)x_i \oplus \alpha_i T_{r(i)}(x_i)$.

The following theorem shows that our algorithm generates approximate solutions of the common fixed point problem under the presence of summable computational errors.

Theorem 3.1 *Assume that* $\{\Delta_i\}_{i=0}^{\infty} \subset [0, \infty)$, $\{\alpha_i\}_{i=0}^{\infty} \subset (0, 1)$,

$$\Delta = \sum_{i=0}^{\infty} \Delta_i < \infty, \tag{3.1}$$

$$x_0 \in B(\theta, M_0) \cap C, \tag{3.2}$$

for each integer $n \geq 0$, $r(n) \in \{1, \ldots, m\}$,

$$d(x_n, T_{r(n)}(x_i)) = \max\{d(x_n, T_s(x_n)) : s = 1, \ldots, m\}, \tag{3.3}$$

$$x_{n+1} \in C \cap B((1 - \alpha_n)x_n \oplus \alpha_n T_{r(n)}(x_n), \Delta_n), \tag{3.4}$$

$\epsilon \in (0, 1)$, n_0 *and* Q *are natural numbers such that*

$$\Delta_i < 2^{-1}\epsilon \tag{3.5}$$

for each integer $i \geq n_0$ *and*

$$\sum_{n=n_0}^{n_0+Q-1} \alpha_n(1 - \alpha_n)$$

$$> 4(4M_0 + 2\Delta)\epsilon^{-1}\eta^{-1}(6M_0 + 2\Delta + 1, \epsilon(6M_0 + 1 + 2\Delta)^{-1}).$$

Then there exists an integer $n \in \{n_0, \ldots, n_0 + Q - 1\}$ *such that*

$$x_n \in F_\epsilon.$$

Proof Proposition 2.5 and (3.1)–(3.4) imply that for each integer $n \geq 0$,

$$d(x_n, \theta) \leq 3M_0 + \Delta. \tag{3.6}$$

Fix a positive number δ such that

$$\delta \le 4^{-1}\epsilon_0 \eta(6M_0 + 2\Delta + 1, \epsilon(6M_0 + 1 + 2\Delta)^{-1})\alpha_i(1 - \alpha_i), \quad i = 0, \ldots, n_0 + Q. \tag{3.7}$$

Assume that the assertion of the theorem does not hold and that $n \in \{n_0, \ldots, n_0 + Q - 1\}$. Then in view of (3.3),

$$d(x_n, T_{r(n)}(x_n)) > \epsilon. \tag{3.8}$$

By assumption (A), Eqs. (3.6)–(3.8), and Lemma 2.4 applied with $S = T_{r(n)}, u = x_n, \alpha = \alpha_n, M = 3M_0 + \Delta, p = p_\delta, \gamma = \epsilon$, and

$$v = (1 - \alpha_n)x_n \oplus \alpha_n T_{r(n)}(x_n)$$

we have

$$d((1 - \alpha_n)x_n \oplus \alpha_n T_{r(n)}(x_n), p_\delta)$$

$$\le d(x_n, p_\delta) - 4^{-1}\epsilon\alpha_n(1 - \alpha_n)\eta(6M_0 + 2\Delta + 1, \epsilon(6M_0 + 2\Delta + 1)^{-1}). \tag{3.9}$$

In view of (3.4) and (3.9),

$$d(x_{n+1}, p_\delta) \le d(x_{n+1}, (1 - \alpha_n)x_n \oplus \alpha_n T_{r(n)}(x_n))$$

$$+ d((1 - \alpha_n)x_n \oplus \alpha_n T_{r(n)}(x_n), p_\delta)$$

$$\le d(x_n, p_\delta) - 4^{-1}\epsilon\alpha_n(1 - \alpha_n)\eta(6M_0 + 2\Delta + 1, \epsilon(6M_0 + 2\Delta + 1)^{-1}) + \Delta_n. \tag{3.10}$$

By assumption (A), (3.6), and (3.10),

$$4M_0 + \Delta \ge d(x_{n_0}, \theta) + d(\theta, p_\delta) \ge d(x_{n_0}, p_\delta)$$

$$= d(x_{n_0}, p_\delta) - d(x_{n_0+Q}, p_\delta)$$

$$\ge \sum_{n=n_0}^{n_0+Q-1} (d(x_n, p_\delta) - d(x_{(n+1)}, p_\delta))$$

$$\ge (\sum_{n=n_0}^{n_0+Q-1} \alpha_n(1 - \alpha_n))4^{-1}\epsilon\eta(6M_0 + 2\Delta + 1, \epsilon(6M_0 + 2\Delta + 1)^{-1})$$

$$- \sum_{n=n_0}^{n_0+Q-1} \Delta_n$$

and

$$\sum_{n=n_0}^{n_0+Q-1} \alpha_n(1-\alpha_n)$$

$$\leq 4^{-1}(4M_0+2\Delta)\epsilon^{-1}\eta^{-1}(6M_0+2\Delta+1, \epsilon(6M_0+2\Delta+1)^{-1}).$$

This contradicts our definition of Q. The contradiction we have reached proves Theorem 3.1.

Theorem 3.1 implies the following result.

Theorem 3.2 *Assume that* $\{\Delta_i\}_{i=0}^{\infty} \subset [0,\infty)$, $\{\alpha_i\}_{i=0}^{\infty} \subset (0,1)$, $\sum_{i=0}^{\infty}\Delta_i < \infty$, $\sum_{n=0}^{\infty}\alpha_n(1-\alpha_n) = \infty$,

$$x_0 \in B(\theta, M_0) \cap C,$$

for each integer $n \geq 0$, $r(n) \in \{1,\dots,m\}$,

$$d(x_n, T_{r(n)}(x_i)) = \max\{d(x_n, T_s(x_n)) : s = 1,\dots,m\}$$

and

$$x_{n+1} \in C \cap B((1-\alpha_n)x_n \oplus \alpha_n T_{r(n)}(x_n), \Delta_n).$$

Then there exists a strictly increasing sequence of natural numbers n_k *such that for each integer* $k \geq 1$, $x_{n_k} \in F_{1/k}$.

3.2   Exact Iterates

The following result describes the behavior of exact iterates of our algorithm.

Theorem 3.3 *Assume that* $\Lambda \in (0, 2^{-1})$,

$$x_0 \in B(\theta, M_0) \cap C, \tag{3.11}$$

$$\{\alpha_i\}_{i=0}^{\infty} \subset (\Lambda, 1-\Lambda), \tag{3.12}$$

for each integer $n \geq 0$, $r(n) \in \{1,\dots,m\}$,

$$d(x_n, T_{r(n)}(x_n)) = \max\{d(x_n, T_s(x_n)) : s = 1,\dots,m\}, \tag{3.13}$$

$$x_{n+1} = (1-\alpha_n)x_n \oplus \alpha_n T_{r(n)}(x_n) \tag{3.14}$$

and that $\epsilon \in (0, 1)$. Then

$$Card(\{n \in \{0, 1, \ldots\} : x_n \notin F_\epsilon\})$$

$$\leq 8M_0\epsilon^{-1}\Lambda^{-2}\eta^{-1}(6M_0 + 1, \epsilon(6M_0 + 1)^{-1}).$$

Proof Proposition 2.5 and (3.11)–(3.14) imply that for each integer $n \geq 0$,

$$d(x_n, \theta) \leq 3M_0. \tag{3.15}$$

Let Q be a natural number,

$$0 < \delta \leq 4^{-1}\epsilon\Lambda^2\eta(6M_0 + 1, \epsilon(6M_0 + 1)^{-1}), \tag{3.16}$$

$$E = \{n \in \{0, 1, \ldots, \} : d(x_n, T_{r(n)}(x_n)) > \epsilon\}, \tag{3.17}$$

$$E_Q = E \cap [0, Q]. \tag{3.18}$$

Assume that

$$n \in E_Q.$$

By assumption (A), the relation above, Eqs. (3.12), (3.14)–(3.18) and Lemma 2.4 applied with $S = T_{r(n)}$, $u = x_n$, $\alpha = \alpha_n$, $M = 3M_0$, $p = p_\delta$, $\gamma = \epsilon$, and $v = x_{n+1}$ we have

$$d(x_{n+1}, p_\delta) \leq d(x_n, p_\delta) - 4^{-1}\epsilon\alpha_n(1 - \alpha_n)\eta(6M_0 + 1, \epsilon(6M_0 + 1)^{-1}). \tag{3.19}$$

Proposition 2.1, assumption (A), and (3.14) imply that for each $n \in \{0, \ldots, Q\}$,

$$d(x_{n+1}, p_\delta) = d((1 - \alpha_n)x_n \oplus \alpha_n T_{r(n)}(x_n), p_\delta)$$

$$\leq (1 - \alpha_n)d(x_n, p_\delta) + \alpha_n d(T_{r(n)}(x_n), p_\delta) \leq d(x_n, p_\delta) + \delta. \tag{3.20}$$

Assumption (A), (3.11), (3.12), (3.18)–(3.20) imply that

$$2M_0 \geq d(x_0, \theta) + d(\theta, p_\delta) \geq d(x_0, p_\delta) \geq d(x_0, p_\delta) - d(x_{Q+1}, p_\delta)$$

$$= \sum_{n=0}^{Q}(d(x_n, p_\delta) - d(x_{n+1}, p_\delta))$$

$$\geq \sum\{d(x_n, p_\delta) - d(x_{n+1}, p_\delta) : n \in E_Q\}$$

$$+ \sum \{d(x_n, p_\delta) - d(x_{n+1}, p_\delta) : \ n \in \{0, \dots, Q\} \setminus E_Q\}$$

$$\geq 4^{-1} \epsilon \Lambda^2 \eta (6M_0 + 1, \epsilon (6M_0 + 1)^{-1}) \mathrm{Card}(E_Q) - \delta(Q + 1),$$

$$4^{-1} \epsilon \Lambda^2 \eta (6M_0 + 1, \epsilon (6M_0 + 1)^{-1}) \mathrm{Card}(E_Q) \leq 2M_0 + \delta(Q + 1).$$

Since δ is any positive number satisfying (3.16), we conclude that

$$\mathrm{Card}(E_Q) \leq 8M_0 \epsilon^{-1} \Lambda^{-2} \eta^{-1} (6M_0 + 1, \epsilon (6M_0 + 1)^{-1}).$$

Since Q is any natural number, we conclude that

$$\mathrm{Card}(E) \leq 8M_0 \epsilon^{-1} \Lambda^{-2} \eta^{-1} (6M_0 + 1, \epsilon (6M_0 + 1)^{-1}).$$

By (3.13) and (3.17), for each $n \in \{0, 1, \dots\} \setminus E$,

$$\epsilon \geq d(x_n, T_{r(n)}(x_n)) \geq d(x_n, T_s(x_n)), \ s = 1, \dots, m$$

and $x_n \in F_\epsilon$. Theorem 3.3 is proved.

3.3   Inexact Iterates with Summable Errors

The following theorem describes the behavior of inexact iterates of our algorithm with summable errors.

Theorem 3.4 *Assume that* $\Lambda \in (0, 2^{-1})$, $\{\Delta_i\}_{i=0}^{\infty} \subset [0, \infty)$,

$$\Delta = \sum_{i=0}^{\infty} \Delta_i < \infty, \tag{3.21}$$

$$\{\alpha_i\}_{i=0}^{\infty} \subset (\Lambda, 1 - \Lambda), \tag{3.22}$$

$$x_0 \in B(\theta, M_0) \cap C, \tag{3.23}$$

for each integer $n \geq 0$, $r(n) \in \{1, \dots, m\}$,

$$d(x_n, T_{r(n)}(x_n)) = \max\{d(x_n, T_s(x_n)) : \ s = 1, \dots, m\}, \tag{3.24}$$

$$x_{n+1} \in B((1 - \alpha_n)x_n \oplus \alpha_n T_{r(n)}(x_n), \Delta_n) \cap C \tag{3.25}$$

and $\epsilon \in (0, 1)$. Then

$$Card(\{n \in \{0, 1, \ldots\} : x_n \notin F_\epsilon\})$$

$$\leq 4(2M_0 + \Delta)\epsilon^{-1}\Lambda^{-2}\eta^{-1}(6M_0 + 2\Delta + 1, \epsilon(6M_0 + 2\Delta + 1)^{-1}).$$

Proof Proposition 2.5 and (3.21), (3.23)–(3.25) imply that for each integer $n \geq 0$,

$$d(x_n, \theta) \leq 3M_0 + \Delta. \tag{3.26}$$

Set

$$E = \{n \in \{0, 1, \ldots\} : d(x_n, T_{r(n)}(x_n)) > \epsilon\}. \tag{3.27}$$

Let Q be a natural number and

$$0 < \delta < 4^{-1}\epsilon\eta(6M_0 + 1 + 2\Delta, \epsilon(6M_0 + 1 + 2\Delta)^{-1})\Lambda^2.$$

Proposition 2.5 and (3.21), (3.23)–(3.25) imply that for each integer $n \geq 0$,

$$d(x_{n+1}, p_\delta) \leq d(x_n, p_\delta) + \delta + \Delta_n. \tag{3.28}$$

Assume that

$$n \in E.$$

By assumption (A), Eqs. (3.22), (3.26), (3.27), the choice of δ, and Lemma 2.4 applied with $S = T_{r(n)}$, $u = x_n$, $\alpha = \alpha_n$, $M = 3M_0 + \Delta$, $p = p_\delta$, $\gamma = \epsilon$, and

$$v = (1 - \alpha_n)x_n \oplus \alpha_n T_{r(n)}(x_n),$$

we have

$$d((1 - \alpha_n)x_n \oplus \alpha_n T_{r(n)}(x_n), p_\delta)$$

$$\leq d(x_n, p_\delta) - 4^{-1}\epsilon\Lambda^2\eta(6M_0 + 2\Delta + 1, \epsilon(6M_0 + 2\Delta + 1)^{-1}). \tag{3.29}$$

In view of (3.25) and (3.29),

$$d(x_{n+1}, p_\delta) \leq d(x_{n+1}, (1 - \alpha_n)x_n \oplus \alpha_n T_{r(n)}(x_n))$$

$$+ d((1 - \alpha_n)x_n \oplus \alpha_n T_{r(n)}(x_n), p_\delta)$$

$$\leq \Delta_n + d(x_n, p_\delta) - 4^{-1}\epsilon\Lambda^2\eta(6M_0 + 2\Delta + 1, \epsilon(6M_0 + 2\Delta + 1)^{-1}). \tag{3.30}$$

Assumption (A) and (3.23), (3.28), (3.30) imply that

$$2M_0 \geq d(x_0, \theta) + d(\theta, p_\delta) \geq d(x_0, p_\delta) \geq d(x_0, p_\delta) - d(x_{Q+1}, p_\delta)$$

$$= \sum_{n=0}^{Q} (d(x_n, p_\delta) - d(x_{n+1}, p_\delta))$$

$$\geq \sum \{d(x_n, p_\delta) - d(x_{n+1}, p_\delta) : \ n \in E \cap [0, Q]\}$$

$$+ \sum \{d(x_n, p_\delta) - d(x_{n+1}, p_\delta) : \ n \in \{0, \ldots, Q\} \setminus E\}$$

$$\geq \sum \{4^{-1} \epsilon \Lambda^2 \eta(6M_0 + 2\Delta + 1, \epsilon(6M_0 + 2\Delta + 1)^{-1}) - \Delta_n : \ n \in E \cap [0, Q]\}$$

$$- \sum \{-\Delta_n - \delta : \ n \in \{0, \ldots, Q\} \setminus E\},$$

$$4^{-1} \epsilon \Lambda^2 \eta(6M_0 + 2\Delta + 1, \epsilon(6M_0 + 2\Delta + 1)^{-1}) \mathrm{Card}(E \cap [0, Q])$$

$$\leq 2M_0 + \Delta + \delta(Q + 1).$$

Since δ is any sufficiently small positive number, we conclude that

$$\mathrm{Card}(E \cap [0, Q]) \leq 4(2M_0 + \Delta)\epsilon^{-1} \Lambda^{-2} \eta^{-1}(6M_0 + 2\Delta + 1, \epsilon(6M_0 + 2\Delta + 1)^{-1}).$$

Since Q is any natural number using (3.24) and (3.27), we conclude that

$$\mathrm{Card}(\{n \in \{0, 1, \ldots\} : \ x_n \notin F_\epsilon\}) \leq \mathrm{Card}(E)$$

$$\leq 4(2M_0 + \Delta)\epsilon^{-1} \Lambda^{-2} \eta^{-1}(6M_0 + 2\Delta + 1, \epsilon(6M_0 + 2\Delta + 1)^{-1}).$$

Theorem 3.4 is proved.

3.4 Inexact Iterates with Nonsummable Errors

The following theorem describes the behavior of inexact iterates of our algorithm with nonsummable errors. It shows that if computational errors are small enough, then our method generates approximate solution belonging to the set $\tilde{F}_\gamma$ with $\gamma > 0$. Our result shows the dependence of γ on our computational errors and provides a number of iterates which should be done in order to obtain this approximate solution.

Theorem 3.5 *Assume that*

$$\Lambda \in (0, 2^{-1}),$$

$\epsilon_0 \in (0, 1),$

$$0 < \delta_0 \leq 8^{-1}\epsilon_0 \Lambda^2 \eta(6M_0 + 3, \epsilon_0(6M_0 + 3)^{-1}), \tag{3.31}$$

$$n_0 = \lfloor 8(2M_0 + 1)\epsilon_0^{-1}\Lambda^{-2}\eta^{-1}(6M_0 + 3, \epsilon_0(6M_0 + 3)^{-1})\rfloor + 1. \tag{3.32}$$

Assume that

$$\{\alpha_i\}_{i=0}^\infty \subset (\Lambda, 1 - \Lambda), \tag{3.33}$$

$$x_0 \in B(\theta, M_0) \cap C, \tag{3.34}$$

for each integer $n \geq 0$, $r(n) \in \{1, \dots, m\}$,

$$d(x_n, T_{r(n)}(x_i)) = \max\{d(x_n, T_s(x_n)) : s = 1, \dots, m\} \tag{3.35}$$

and that

$$x_{n+1} \in B((1 - \alpha_n)x_n \oplus \alpha_n T_{r(n)}(x_n), \delta_0). \tag{3.36}$$

Then there exists an integer $q \in [1, n_0]$ such that $d(x_i, \theta) \leq 3M_0 + 1$, $i = 0, \dots, q$ and $x_q \in F_{\epsilon_0}$.

Proof Assume that s is a natural number and that for each $k \in [1, s]$,

$$d(x_k, T_{r(k)}(x_k)) > \epsilon_0. \tag{3.37}$$

Fix

$$\delta \in (0, 2^{-1}\delta_0). \tag{3.38}$$

Assumption (A), Proposition 2.1, and (3.31), (3.34), (3.36), and (3.38) imply that

$$d(x_1, p_\delta) \leq d(x_1, (1 - \alpha_0)x_0 \oplus \alpha_0 T_{r(0)}(x_0))$$

$$+ d((1 - \alpha_0)x_0 \oplus \alpha_0 T_{r(0)}(x_0), p_\delta)$$

$$\leq \delta_0 + (1 - \alpha_0)d(x_0, p_\delta) + \alpha_0 d(T_{r(0)}(x_0), p_\delta)$$

$$\leq \delta_0 + d(x_0, p_\delta) + \delta \leq 2M_0 + 1. \tag{3.39}$$

Assume that $k \in [1, s]$ as an integer and

$$d(x_k, p_\delta) \leq 2M_0 + 1. \tag{3.40}$$

(In view of (3.39), inequality (3.40) holds for $k = 1$.) Then

$$d(x_k, \theta) \leq 3M_0 + 1.$$

By assumption (A), Eqs. (3.31), (3.33), (3.37), (3.38), and Lemma 2.4 applied with $S = T_{r(k)}$, $u = x_k$, $\alpha = \alpha_k$, $M = 3M_0 + 1$, $p = p_\delta$, $\gamma = \epsilon_0$, and

$$v = (1 - \alpha_k)x_k \oplus \alpha_k T_{r(k)}(x_k),$$

we have

$$d((1 - \alpha_k)x_k \oplus \alpha_k T_{r(k)}(x_k), p_\delta)$$

$$\leq d(x_k, p_\delta) - 4^{-1}\epsilon_0 \Lambda^2 \eta(6M_0 + 3, \epsilon_0(6M_0 + 3)^{-1}).$$

Together with (3.31) and (3.36), this implies that

$$d(x_{k+1}, p_\delta) \leq d(x_{k+1}, (1 - \alpha_k)x_k \oplus \alpha_k T_{r(k)}(x_k))$$

$$+ d((1 - \alpha_k)x_k \oplus \alpha_k T_{r(k)}(x_k), p_\delta)$$

$$\leq \delta_0 + d(x_k, p_\delta) - 4^{-1}\epsilon_0 \Lambda^2 \eta(6M_0 + 3, \epsilon_0(6M_0 + 3)^{-1})$$

$$\leq d(x_k, p_\delta) - 8^{-1}\epsilon_0 \Lambda^2 \eta(6M_0 + 3, \epsilon_0(6M_0 + 3)^{-1}).$$

By induction we obtain that for all $k = 1, \ldots, s + 1$,

$$d(x_k, p_\delta) \leq 2M_0 + 1, \tag{3.41}$$

and for all $k = 1, \ldots, s$,

$$d(x_{k+1}, p_\delta) \leq d(x_k, p_\delta) - 8^{-1}\epsilon_0 \Lambda^2 \eta(6M_0 + 3, \epsilon_0(6M_0 + 3)^{-1}). \tag{3.42}$$

It follows from (3.39) and (3.42) that

$$8s\epsilon_0 \Lambda^2 \eta(6M_0 + 3, \epsilon_0(6M_0 + 3)^{-1}) \leq \sum_{k=1}^{s}(d(x_k, p_\delta) - d(x_{(k+1)}, p_\delta))$$

$$\leq d(x_1, p_\delta) \leq 2M_0 + 1,$$

$$s \leq 8(2M_0 + 1)\epsilon_0^{-1} \Lambda^{-2} \eta^{-1}(6M_0 + 3, \epsilon_0(6M_0 + 3)^{-1}).$$

Thus we have shown that the following property holds:

(P) If an integer $s \geq 1$ and for each $k \in \{1, \ldots, s\}$ (3.37) holds, then

$$s \leq 8(2M_0 + 1)\epsilon_0^{-1}\Lambda^{-2}\eta^{-1}(6M_0 + 3, \epsilon_0(6M_0 + 3)^{-1}),$$

$$d(x_j, p_\delta) \leq 2M_0 + 1, \quad j = 1, \ldots, (s + 1).$$

Property (P) and (3.35) imply that there exists an integer $q \in [1, n_0]$ such that

$$d(x_q, T_{r(q)}(x_q)) \leq \epsilon_0, \quad x_q \in F_{\epsilon_0},$$

if an integer k satisfies $1 \leq k < q$, then

$$d(x_k, T_{r(k)}(x_q)) > \epsilon_0,$$

$$d(x_k, p_\delta) \leq 2M_0 + 1, \quad d(x_k, \theta) \leq 3M_0 + 1, \quad k = 1, \ldots, q.$$

This completes the proof of Theorem 3.5.

Theorem 3.5 implies the following result.

Theorem 3.6 *Assume that $\bar{\epsilon} \in (0, 1)$, $\tilde{F}_{\bar{\epsilon}} \subset B(\theta, M_0)$, $0 < \epsilon_0 \leq \bar{\epsilon}$, $\Lambda \in (0, 2^{-1})$,*

$$0 < \delta_0 \leq 8^{-1}\epsilon_0\Lambda^2\eta(6M_0 + 3, \epsilon_0(6M_0 + 3)^{-1}),$$

$$n_0 = \lfloor 8(2M_0 + 1)\epsilon_0^{-1}\Lambda^{-2}\eta^{-1}(6M_0 + 3, \epsilon_0(6M_0 + 3)^{-1})\rfloor + 1.$$

Assume that

$$\{\alpha_i\}_{i=0}^{\infty} \subset (\Lambda, 1 - \Lambda),$$

$$x_0 \in B(\theta, M_0) \cap C,$$

for each integer $n \geq 0$, $r(n) \in \{1, \ldots, m\}$,

$$d(x_n, T_{r(n)}(x_i)) = \max\{d(x_n, T_s(x_n)) : s = 1, \ldots, m\},$$

$$x_{n+1} \in B((1 - \alpha_n)x_n \oplus \alpha_n T_{r(n)}(x_n), \delta_0).$$

Then there exists a strictly increasing sequence of integers $\{n_k\}_{k=1}^{\infty}$ such that $1 \leq n_1 \leq n_0$ and that for each integer $k \geq 1$, $1 \leq n_{k+1} - n_k \leq n_0$ and $x_{n_k} \in F_{\epsilon_0}$.

Chapter 4
Set-Valued Inclusions

Abstract In this chapter we use Krasnoselskii-Mann iterations for solving set-valued inclusions in a W-hyperbolic space. Our main goal is to obtain an approximate solution of the problem in the presence of computational errors. We show that the iterative method generates a good approximate solution, if the sequence of computational errors is bounded from above by a constant. Moreover, for a known computational error, we find out what an approximate solution can be obtained and how many iterates one needs for this.

4.1 Preliminaries

Assume that $X = (X, d, W)$ is a W-hyperbolic space with a structure (X, η), where $\eta : (0, \infty) \times (0, 2] \to (0, 1]$ is its modulus of uniform convexity.

We continue to use the notation and definitions introduced in Sects. 2.1–2.3. In particular, we assume that $C \subset X$ is a nonempty convex set,

$$M_0 > 0, \theta \in C.$$

We assume that $T : C \to 2^C \setminus \{\emptyset\}$, $T(x)$ is bounded for each $x \in X$ and the following assumption holds.

(B)   For each $\delta \in (0, 1)$ there exists $p_\delta \in B(\theta, M_0) \cap C$ such that for each $x \in C$ and each $y \in T(x)$,

$$d(y, p_\delta) \leq d(x, p_\delta) + \delta.$$

For each $\delta \in (0, 1)$ let p_δ be as guaranteed by (B). Clearly, for each $\delta \in (0, 1)$, $T(p_\delta) \subset B(p_\delta)$.

Now we describe our algorithm.

Initialization: Let $\delta > 0$. Choose

$$x \in C \cap B(\theta, M_0)$$

and set

$$x_0 = x.$$

Iterative step: For each integer $i \geq 0$, choose $\alpha_i \in (0, 1)$ and $y_i \in T(x_i)$ such that

$$d(x_i, y_i) \geq \sup\{d(x_i, y) :\ y \in T(x_i)\} - \delta$$

and set

$$x_{i+1} = (1 - \alpha_i)x_i \oplus \alpha_i y_i.$$

The following lemma is an important ingredient in our study in this chapter.

Lemma 4.1 *Assume that* $M > 0$, $\alpha, \gamma \in (0, 1)$,

$$0 < \delta \leq 4^{-1}\gamma\alpha(1 - \alpha)\eta(2M + 1, \gamma(2M + 1)^{-1}), \tag{4.1}$$

$$p \in B(\theta, M) \cap C, \tag{4.2}$$

$$d(y, p) \leq d(x, p) + \delta \text{ for each } x \in C \text{ and each } y \in T(x), \tag{4.3}$$

$$u \in B(\theta, M) \cap C, \tag{4.4}$$

$$v \in T(u), \tag{4.5}$$

$$d(u, v) \geq \gamma. \tag{4.6}$$

Then

$$d((1 - \alpha)u \oplus \alpha v, p) \leq d(u, p) + 1 \leq 2M + 1,$$

$$d((1 - \alpha)u \oplus \alpha v, p) \leq d(u, p) - 4^{-1}\gamma\alpha(1 - \alpha)\eta(2M + 1, \gamma(2M + 1)^{-1}).$$

Proof Proposition 2.1 and Eqs. (4.2)–(4.5) imply that

$$d((1 - \alpha)u \oplus \alpha v, p) \leq (1 - \alpha)d(u, p) + \alpha d(v, p)$$

$$\leq (1 - \alpha)d(u, p) + \alpha d(u, p) + \delta$$

$$\leq d(u, p) + 1 \leq d(u, \theta) + d(\theta, p) + 1 \leq 2M + 1,$$

$$d((1 - \alpha)u \oplus \alpha v), \theta) \leq 3M + 1. \tag{4.7}$$

By (4.1), (4.3), (4.5), and (4.6),

$$\gamma \leq d(u, v) \leq d(u, p) + d(p, v) \leq d(u, p) + d(p, u) + \delta,$$

$$d(u, p) \geq 2^{-1}(\gamma - \delta) \geq 4^{-1}\gamma. \tag{4.8}$$

In view of (4.2), (4.4), and (4.6),

$$d(u, v) \geq \gamma \geq \gamma(\delta + d(u, p))(2M + 1)^{-1}.$$

By (4.3) and (4.5),

$$d(p, v) \leq \delta + d(p, u).$$

Equations above, (4.1), (4.2), (4.4), Lemma 2.2 applied with

$$x = u, \ y = v, \ a = p, \ r = \delta + d(u, p),$$

$$s = 2M + 1, \ \epsilon = \gamma(2M + 1)^{-1}, \ \lambda = \alpha$$

and Eqs. (4.1), (4.8) imply that

$$d((1 - \alpha)u \oplus \alpha v), p)$$

$$\leq (1 - 2\alpha(1 - \alpha)\eta(2M + 1, \gamma(2M + 1)^{-1})(\delta + d(u, p))$$

$$\leq d(u, p) + \delta - 2\alpha(1 - \alpha)\eta(2M + 1, \gamma(2M + 1)^{-1})(4^{-1}\gamma)$$

$$\leq d(u, p) - 4^{-1}\gamma\alpha(1 - \alpha)\eta(2M + 1, \gamma(2M + 1)^{-1}).$$

Lemma 4.1 is proved.

4.2 Inexact Iterates with Summable Errors

The results of this section describe the behavior of inexact iterates of our algorithm with summable errors.

Proposition 4.2 *Assume that* $\{r_i\}_{i=0}^{\infty} \subset [0, \infty),$

$$r = \sum_{i=0}^{\infty} r_i < \infty,$$

$\{\alpha_i\}_{i=0}^{\infty} \subset (0, 1),$

$$x_0 \in B(\theta, M_0) \cap C, \tag{4.9}$$

for each integer $n \geq 0$,

$$y \in T(x_n), \tag{4.10}$$

$$x_{n+1} \in C \cap B((1 - \alpha_n)x_n \oplus \alpha_n y_n, r_n). \tag{4.11}$$

Then for each integer $n \geq 0$ and each $\delta \in (0, 1)$,

$$d(x_{n+1}, p_\delta) \leq d(x_n, p_\delta) + \delta + r_n,$$

$$d(x_n, \theta) \leq 3M_0 + r, \ \ d(y_n, \theta) \leq 3M_0 + \Delta.$$

Proof Let $\delta \in (0, 1)$ and $n \geq 0$ be an integer. Assumption (B) and Eqs. (4.10) and (4.11) imply that

$$d(y_n, p_\delta) \leq d(x_n, p_\delta) + \delta. \tag{4.12}$$

Proposition 2.1 and (4.11), (4.12) imply that

$$d(x_{n+1}, p_\delta) \leq d(x_{n+1}, (1 - \alpha_n)x_n \oplus \alpha_n y_n)$$

$$+ d((1 - \alpha_n)x_n \oplus \alpha_n y_n, p_\delta)$$

$$\leq r_n + (1 - \alpha_n)d(x_n, p_\delta) + \alpha_n d(y_n, p_\delta)$$

$$\leq r_n + \delta + d(x_n, p_\delta).$$

Assumption (B), (4.9), (4.12), and the relation above imply that for each integer $n \geq 0$,

$$d(x_n, p_\delta) \leq d(x_0, p_\delta) + n\delta + \sum_{i=0}^{n} r_i \leq 2M_0 + r + n\delta,$$

$$d(y_n, p_\delta) \leq 2M_0 + r + (n + 1)\delta,$$

$$d(x_n, \theta) \leq 3M_0 + n\delta + r,$$

$$d(y_n, \theta) \leq 3M_0 + (n + 1)\delta + r.$$

Since δ is any element of the interval $(0, 1)$, we conclude that for each integer $n \geq 0$,

$$d(x_n, \theta) \leq 3M_0 + r, \quad d(y_n, \theta) \leq 3M_0 + r.$$

Proposition 4.2 is proved.

Theorem 4.3 *Assume that* $\{\Delta_i\}_{i=0}^{\infty} \subset [0, \infty)$, $\{r_i\}_{i=0}^{\infty} \subset [0, \infty)$, $\{\alpha_i\}_{i=0}^{\infty} \subset (0, 1)$,

$$\lim_{i \to \infty} \Delta_i = 0, \tag{4.13}$$

$$r = \sum_{i=0}^{\infty} r_i < \infty, \tag{4.14}$$

$$x_0 \in B(\theta, M_0) \cap C, \tag{4.15}$$

for each integer $n \geq 0$,

$$y_n \in T(x_n), \tag{4.16}$$

$$d(x_n, y_n) \geq \sup\{d(x_n, \xi) : \xi \in T(x_n)\} - \Delta_n, \tag{4.17}$$

$$x_{n+1} \in C \cap B((1 - \alpha_n)x_n \oplus \alpha_n y_n, r_n), \tag{4.18}$$

$\epsilon \in (0, 1)$, n_0, *and* Q *are natural numbers such that*

$$\Delta_n < 2^{-1}\epsilon \tag{4.19}$$

for each integer $n \geq n_0$ *and*

$$\sum_{n=n_0}^{n_0+Q-1} \alpha_n(1 - \alpha_n)$$

$$> 8(4M_0 + 2r)\epsilon^{-1}\eta^{-1}(6M_0 + 2r + 1, 2^{-1}\epsilon(6M_0 + 1 + 2r)^{-1}). \tag{4.20}$$

Then there exists an integer $n \in \{n_0, \ldots, n_0 + Q - 1\}$ *such that*

$$T(x_n) \in B(x_n, \epsilon).$$

Proof Proposition 4.2 and (4.14)–(4.18) imply that for each integer $n \geq 0$,

$$d(x_n, \theta), \ d(y_n, \theta) \leq 3M_0 + r. \tag{4.21}$$

Assume that the assertion of the theorem does not hold. Then for each $n \in \{n_0, \ldots, n_0 + Q - 1\}$,

$$T(x_n) \setminus B(x_n, \epsilon) \neq \emptyset. \qquad (4.22)$$

Fix a positive number δ such that

$$\delta < 8^{-1} \epsilon \eta (6M_0 + 2r + 1, 2^{-1} \epsilon (6M_0 + 1 + 2r)^{-1}) \alpha_i (1 - \alpha_i), \quad i = 0, \ldots, n_0 + Q. \qquad (4.23)$$

Let

$$n \in \{n_0, \ldots, n_0 + Q - 1\}. \qquad (4.24)$$

Proposition 2.1, assumption (B), (4.16), and (4.18) imply that

$$d(x_{n+1}, p_\delta) \leq d(x_{n+1}, (1 - \alpha_n)x_n \oplus \alpha_n y_n)$$

$$+ d((1 - \alpha_n)x_n \oplus \alpha_n y, p_\delta)$$

$$\leq r_n + (1 - \alpha_n)d(x_n, p_\delta) + (1 - \alpha_n)d(y_n, p_\delta) \leq r_n + d(x_n, p_\delta) + \delta. \qquad (4.25)$$

In view of (4.17), (4.19), and (4.22),

$$d(x_n, y_n) \geq \epsilon - \Delta_n \geq \epsilon/2. \qquad (4.26)$$

By assumption (B), Eqs. (4.16), (4.21), (4.23), (4.26), and Lemma 4.2 applied with $u = x_n$, $\alpha = \alpha_n$, $M = 3M_0 + r$, $p = p_\delta$, $\gamma = \epsilon/2$, $v = y_n$, we have

$$d((1 - \alpha_n)x_n \oplus \alpha_n y_n, p_\delta)$$

$$\leq d(x_n, p_\delta) - 8^{-1} \epsilon \alpha_n (1 - \alpha_n) \eta (6M_0 + 2r + 1, 2^{-1} \epsilon (6M_0 + 2r + 1)^{-1}).$$

Together with (4.18), this implies that

$$d(x_{n+1}, p_\delta) \leq d(x_{n+1}, (1 - \alpha_n)x_n \oplus \alpha_n y_n)$$

$$+ d((1 - \alpha_n)x_n \oplus \alpha_n y, p_\delta)$$

$$\leq d(x_n, p_\delta) - 8^{-1} \epsilon \alpha_n (1 - \alpha_n) \eta (6M_0 + 2r + 1, 2^{-1} \epsilon (6M_0 + 2r + 1)^{-1}) + r_n.$$

Thus for each $n \in \{n_0, \ldots, n_0 + Q - 1\}$,

$$d(x_{n+1}, p_\delta)$$

$$\leq d(x_n, p_\delta) - 8^{-1}\epsilon\alpha_n(1 - \alpha_n)\eta(6M_0 + 2r + 1, 2^{-1}\epsilon(6M_0 + 2r + 1)^{-1}) + r_n. \tag{4.27}$$

By assumption (B), (4.14), (4.21), and (4.27),

$$4M_0 + r \geq d(x_{n_0}, \theta) + d(\theta, p_\delta) \geq d(x_{n_0}, p_\delta)$$

$$= d(x_{n_0}, p_\delta) - d(x_{n_0+Q}, p_\delta)$$

$$= \sum_{n=n_0}^{n_0+Q-1} (d(x_n, p_\delta) - d(x_{n+1}, p_\delta))$$

$$\geq \left(\sum_{n=n_0}^{n_0+Q-1} \alpha_n(1 - \alpha_n) \right) 8^{-1}\epsilon\eta(6M_0 + 2r + 1, 2^{-1}\epsilon(6M_0 + 2r + 1)^{-1})$$

$$- \sum_{n=n_0}^{n_0+Q-1} r_n$$

and

$$\sum_{n=n_0}^{n_0+Q-1} \alpha_n(1 - \alpha_n)$$

$$\leq 8(4M_0 + 2r)\epsilon^{-1}\eta^{-1}(6M_0 + 2r + 1, 2^{-1}\epsilon(6M_0 + 2r + 1)^{-1}).$$

This contradicts (4.20). The contradiction we have reached proves Theorem 4.3.

Theorem 4.3 implies the following result.

Theorem 4.4 *Assume that* $\{\Delta_i\}_{i=0}^{\infty} \subset (0, \infty)$, $\{r_i\}_{i=0}^{\infty} \subset [0, \infty)$, $\{\alpha_i\}_{i=0}^{\infty} \subset (0, 1)$, $\lim_{i \to \infty} \Delta_i = 0$, $\sum_{i=0}^{\infty} r_i < \infty$,

$$\sum_{n=0}^{\infty} \alpha_n(1 - \alpha_n) = \infty,$$

$$x_0 \in B(\theta, M_0) \cap C,$$

for each integer $n \geq 0$, $y_n \in T(x_n)$,

$$d(x_n, y_n) \geq \sup\{d(x_n, \xi) : \xi \in T(x_n)\} - \Delta_n$$

and that

$$x_{n+1} \in C \cap B((1 - \alpha_n)x_n \oplus \alpha_n y_n, r_n).$$

Then there exists a strictly increasing sequence of natural numbers n_k such that for each integer $k \geq 1$, $T(x_{n_k}) \subset B(x_{n_k}, 1/k)$.

Theorem 4.5 *Assume that $\Lambda \in (0, 2^{-1})$, $\{\Delta_i\}_{i=0}^{\infty} \subset [0, \infty)$, $\{r_i\}_{i=0}^{\infty} \subset [0, \infty)$, $\lim_{i \to \infty} \Delta_i = 0$,*

$$r = \sum_{i=0}^{\infty} r_i < \infty, \quad \{\alpha_i\}_{i=0}^{\infty} \subset (\Lambda, 1 - \Lambda), \tag{4.28}$$

$$x_0 \in B(\theta, M_0) \cap C, \tag{4.29}$$

for each integer $n \geq 0$,

$$y_n \in T(x_n), \tag{4.30}$$

$$d(x_n, y_n) \geq \sup\{d(x_n, \xi) : \xi \in T(x_n)\} - \Delta_n, \tag{4.31}$$

$$x_{n+1} \in C \cap B((1 - \alpha_n)x_n \oplus \alpha_n y_n, r_n), \tag{4.32}$$

and $\epsilon \in (0, 1)$. Let n_0 be a natural number such that for each integer $n \geq n_0$,

$$\Delta_i \leq \epsilon/2. \tag{4.33}$$

Then

$$Card(\{n \in \{0, 1, \ldots\} : T(x_n) \setminus B(x_n, \epsilon) \neq \emptyset\})$$

$$\leq n_0 + 8(4M_0 + 2r)\epsilon^{-1}\Lambda^{-2}\eta^{-1}(6M_0 + 2r + 1, 2^{-1}\epsilon(6M_0 + 2r + 1)^{-1}).$$

Proof Proposition 4.2 and (4.28)–(4.32) imply that for each integer $n \geq 0$,

$$d(x_n, \theta) \leq 3M_0 + r, \quad d(x_n, \theta) \leq 3M_0 + r. \tag{4.34}$$

Set

$$E_0 = \{n \in \{0, 1, \ldots\} : T(x_n) \setminus B(x_n, \epsilon) \neq \emptyset\}. \tag{4.35}$$

Let $Q > n_0$ be a natural number and

$$0 < \delta < 8^{-1} \epsilon \eta (6M_0 + 1 + 2r, 2^{-1} \epsilon (6M_0 + 1 + 2r)^{-1}) \Lambda^2. \tag{4.36}$$

Assume that

$$n \in E_0 \cap [n_0, Q]. \tag{4.37}$$

In view of (4.31), (4.33), (4.35), and (4.37),

$$d(x_n, y_n) \geq \sup\{d(x_n, \xi) : \xi \in T(x_n)\} - \Delta_n \geq \epsilon - \Delta_n \geq 2^{-1} \epsilon. \tag{4.38}$$

Proposition 4.2 and (4.28)–(4.32) imply that

$$d(x_{k+1}, p_\delta) \leq d(x_k, p_\delta) + \delta + r_k \tag{4.39}$$

for each integer $k \geq 0$. By assumption (B), Eqs. (4.28), (4.30), (4.34)), (4.36), (4.38) and Lemma 4.1 applied with $u = x_n$, $\alpha = \alpha_n$, $M = 3M_0 + r$, $p = p_\delta$, $\gamma = \epsilon/2$, and $v = y_n$, we have

$$d((1 - \alpha_n) x_n \oplus \alpha_n y_n, p_\delta)$$

$$\leq d(x_n, p_\delta) - 8^{-1} \epsilon \alpha_n (1 - \alpha_n) \eta (6M_0 + 2r + 1, 2^{-1} \epsilon (6M_0 + 2r + 1)^{-1}).$$

Together with (4.28) and (4.32), this implies that

$$d(x_{n+1}, p_\delta) \leq d(x_{n+1}, (1 - \alpha_n) x_n \oplus \alpha_n y_n)$$

$$+ d((1 - \alpha_n) x_n \oplus \alpha_n y_n, p_\delta)$$

$$\leq r_n + d(x_n, p_\delta) - 8^{-1} \epsilon \Lambda^2 \eta (6M_0 + 2r + 1, 2^{-1} \epsilon (6M_0 + 2r + 1)^{-1}). \tag{4.40}$$

Assumption (B) and (4.34), (4.39), (4.40) imply that

$$4M_0 + r \geq d(x_{n_0}, p_\delta) \geq d(x_{n_0}, p_\delta) - d(x_{Q+1}, p_\delta)$$

$$= \sum_{n=n_0}^{Q} (d(x_n, p_\delta) - d(x_{n+1}, p_\delta))$$

$$= \sum \{d(x_n, p_\delta) - d(x_{n+1}, p_\delta) : n \in E_0 \cap [n_0, Q]\}$$

$$+ \sum \{d(x_n, p_\delta) - d(x_{n+1}, p_\delta) : n \in \{n_0, \ldots, Q\} \setminus E_0\}$$

$$\geq 8^{-1}\epsilon\Lambda^2\eta(6M_0 + 2r + 1, 2^{-1}\epsilon(6M_0 + 2r + 1)^{-1})\mathrm{Card}(E_0 \cap [n_0, \varrho\,])$$

$$-\sum\{r_n : n \in \{n_0, \dots, Q\} \cap E_0\} - \sum\{r_n : n \in \{n_0, \dots, Q\} \setminus E_0\} - \delta(Q+1),$$

$$\mathrm{Card}(E_0 \cap [n_0, Q])$$

$$\leq 8(4M_0 + 2r + \delta(Q+1))\epsilon^{-1}\Lambda^{-2}\eta^{-1}(6M_0 + 2r + 1, 2^{-1}\epsilon(6M_0 + 2r + 1)^{-1}).$$

Since δ is any sufficiently small positive number, we conclude that

$$\mathrm{Card}(E_0 \cap [n_0, Q])$$

$$\leq 8(4M_0 + 2r)\epsilon^{-1}\Lambda^{-2}\eta^{-1}(6M_0 + 2r + 1, 2^{-1}\epsilon(6M_0 + 2r + 1)^{-1}).$$

Since Q is any large enough natural number, we conclude that

$$\mathrm{Card}(E_0) \leq n_0 + 8(4M_0 + 2r)\epsilon^{-1}\Lambda^{-2}\eta^{-1}(6M_0 + 2r + 1, 2^{-1}\epsilon(6M_0 + 2r + 1)^{-1}).$$

Theorem 4.5 is proved.

4.3   Inexact Iterates with Nonsummable Errors

The following theorem describes the behavior of inexact iterates of our algorithm with nonsummable errors. It shows that if computational errors are small enough, then our method generates approximate solution x such that $T(x) \subset B(x, \gamma)$ with $\gamma > 0$. Our result shows the dependence of γ on our computational errors and provides a number of iterates which should be done in order to obtain this approximate solution.

Theorem 4.6 *Assume that* $\Lambda \in (0, 2^{-1})$, $\epsilon_0 \in (0, 1/2)$,

$$0 < \delta_0 \leq 16^{-1}\epsilon_0\Lambda^2\eta(6M_0 + 3, 2^{-1}\epsilon_0(6M_0 + 3)^{-1}), \tag{4.41}$$

$$n_0 = \lfloor 16(2M_0 + 1)\epsilon_0^{-1}\Lambda^{-2}\eta^{-1}(6M_0 + 3, 2^{-1}\epsilon_0(6M_0 + 3)^{-1})\rfloor + 1. \tag{4.42}$$

Assume that

$$\{\alpha_i\}_{i=0}^{\infty} \subset (\Lambda, 1 - \Lambda), \tag{4.43}$$

$$x_0 \in B(\theta, M_0) \cap C, \tag{4.44}$$

for each integer n ≥ 0,

$$y_n \in T(x_n), \tag{4.45}$$

$$d(x_n, y_n) \geq \sup\{d(x_n, \xi) : \xi \in T(x_n)\} - \epsilon_0/2, \tag{4.46}$$

$$x_{n+1} \in C \cap B((1 - \alpha_n)x_n \oplus \alpha_n y_n, \delta_0). \tag{4.47}$$

Then there exists an integer $q \in [1, n_0]$ such that

$$d(x_i, \theta) \leq 3M_0 + 1, \ i = 0, \ldots, q,$$

$$T(x_q) \subset B(x_q, \epsilon_0).$$

Proof Assume that s is a natural number and that for each $k \in [1, s]$,

$$T(x_k) \setminus B(x_k, \epsilon_0) \neq \emptyset. \tag{4.48}$$

Fix

$$\delta \in (0, \delta_0). \tag{4.49}$$

In view of (4.46) and (4.48), for each $k \in \{1, \ldots, s\}$,

$$d(x_k, y_k) \geq \sup\{d(x_k, \xi) : \xi \in T(x_k)\} - \epsilon_0/2 \geq \epsilon_0/2. \tag{4.50}$$

Assumption (B), Proposition 2.1, and (4.44), (4.45), and (4.47) imply that

$$d(y_0, p_\delta) \leq d(x_0, p_\delta) + \delta,$$

$$d(x_1, p_\delta) \leq d(x_1, (1 - \alpha_0)x_0 \oplus \alpha_0 y_0)$$

$$+ d((1 - \alpha_0)x_0 \oplus \alpha_0 y_0, p_\delta)$$

$$\leq \delta_0 + (1 - \alpha_0)d(x_0, p_\delta) + \alpha_0 d(y_0, p_\delta)$$

$$\leq \delta_0 + d(x_0, p_\delta) + \delta \leq 2\delta_0 + d(x_0, p_\delta) \leq 2M_0 + 1. \tag{4.51}$$

Assume that $k \in [1, s]$ as an integer and

$$d(x_k, p_\delta) \leq 2M_0 + 1. \tag{4.52}$$

(In view of (4.51), inequality (4.52) holds for $k = 1$.) By assumption (B), Eqs. (4.41), (4.45), (4.49), (4.50), (4.52), and Lemma 4.1 applied with $u = x_k$, $\alpha = \alpha_k$, $M = 3M_0 + 1$, $p = p_\delta$, $\gamma = \epsilon_0/2$, and $v = y_k$, we have

$$d((1 - \alpha_k)x_k \oplus \alpha_k y_k, p_\delta)$$

$$\leq d(x_k, p_\delta) - 8^{-1}\epsilon_0 \alpha_k (1 - \alpha_k)\eta(6M_0 + 3, 2^{-1}\epsilon_0(6M_0 + 3)^{-1}).$$

Together with (4.41) and (4.47), this implies that

$$d(x_{k+1}, p_\delta) \leq d(x_{k+1}, (1 - \alpha_k)x_k \oplus \alpha_k y_k)$$

$$+ d((1 - \alpha_k)x_k \oplus \alpha_k y_k, p_\delta)$$

$$\leq \delta_0 + d(x_k, p_\delta) - 8^{-1}\epsilon_0 \Lambda^2 \eta(6M_0 + 3, 2^{-1}\epsilon_0(6M_0 + 3)^{-1})$$

$$\leq d(x_k, p_\delta) - 16^{-1}\epsilon_0 \Lambda^2 \eta(6M_0 + 3, 2^{-1}\epsilon_0(6M_0 + 3)^{-1}). \qquad (4.53)$$

By induction we obtain that for all $k = 1, \ldots, s + 1$,

$$d(x_k, p_\delta) \leq 2M_0 + 1,$$

and for all $k = 1, \ldots, s$, Eq. (4.53) holds. It follows from (4.51) and (4.53) that

$$2M_0 + 1 \geq d(x_1, p_\delta) \geq d(x_1, p_\delta) - d(x_{s+1}, p_\delta)$$

$$= \sum_{k=1}^{s}(d(x_k, p_\delta) - d(x_{(k+1)}, p_\delta))$$

$$\geq 16^{-1}s\epsilon_0 \Lambda^2 \eta(6M_0 + 3, 2^{-1}\epsilon_0(6M_0 + 3)^{-1}),$$

$$s \leq 16(2M_0 + 1)\epsilon_0^{-1}\Lambda^{-2}\eta^{-1}(6M_0 + 3, 2^{-1}\epsilon_0(6M_0 + 3)^{-1}).$$

Thus we have shown that the following property holds:
    (P1) If an integer $s \geq 1$ and for each $k \in \{1, \ldots, s\}$ (4.48) holds, then

$$s \leq 16(2M_0 + 1)\epsilon_0^{-1}\Lambda^{-2}\eta^{-1}(6M_0 + 3, 2^{-1}\epsilon_0(6M_0 + 3)^{-1}),$$

$$d(x_j, p_\delta) \leq 2M_0 + 1, \quad j = 1, \ldots, s + 1.$$

Property (P1) implies that there exists $q \in \{1, \ldots, n_0\}$ such that

$$T(x_q) \subset B(x_q, \epsilon_0),$$

and if an integer k satisfies $1 \le k < q$, then

$$T(x_k) \setminus B(x_k, \epsilon) \neq \emptyset,$$

$$d(x_k, p_\delta) \le 2M_0 + 1, \ \ d(x_k, \theta) \le 3M_0 + 1, \ k = 1, \dots, q.$$

This completes the proof of Theorem 4.5.

Theorem 4.5 implies the following result.

Theorem 4.7 *Assume that* $\bar{\epsilon} \in (0, 1/2)$, $\tilde{F}_{\bar{\epsilon}} \subset B(\theta, M_0)$, $0 < \epsilon_0 < \bar{\epsilon}$, $\Lambda \in$
$(0, 2^{-1})$,

$$0 < \delta_0 \le 16^{-1} \epsilon_0 \Lambda^2 \eta(6M_0 + 3, 2^{-1}\epsilon_0(6M_0 + 3)^{-1}),$$

$$n_0 = \lfloor 16(2M_0 + 1)\epsilon_0^{-1} \Lambda^{-2} \eta^{-1}(6M_0 + 3, 2^{-1}\epsilon_0(6M_0 + 3)^{-1}) \rfloor + 1.$$

Assume that

$$\{\alpha_i\}_{i=0}^{\infty} \subset (\Lambda, 1 - \Lambda),$$

$$x_0 \in B(\theta, M_0) \cap C,$$

for each integer $n \ge 0$,

$$y_n \in T(x_n),$$

$$d(x_n, y_n) \ge \sup\{d(x_n, \xi) : \xi \in T(x_n)\} - \epsilon_0/2,$$

$$x_{n+1} \in C \cap B((1 - \alpha_n)x_n \oplus \alpha_n y_n, \delta_0).$$

Then there exists a strictly increasing sequence of integers $\{n_k\}_{k=1}^{\infty}$ *such that*

$$1 \le n_1 \le n_0$$

and for each integer $k \ge 1$

$$1 \le n_{k+1} - n_k \le n_0, \ T(x_{n_k}) \subset B(x_{n_k}, \epsilon_0).$$

Chapter 5
The Cimmino Algorithm in a Normed Space

Abstract In this chapter we study the convergence of the Cimmino algorithm for solving common fixed point problems in a normed space using Krasnoselskii-Mann iterations. Our main goal is to obtain an approximate solution of the problem in the presence of computational errors. We show that the iterative method generates a good approximate solution, if the sequence of computational errors is bounded from above by a constant. Moreover, for a known computational error, we find out what an approximate solution can be obtained and how many iterates one needs for this.

5.1 Cimmino Iterates

In this chapter we assume that W-hyperbolic space $X = (X, d, W)$ equipped with the structure (X, η), where $\eta : (0, \infty) \times (0, 2] \to (0, 1]$ is the modulus of uniform convexity, is also a normed space $(X, \| \cdot \|)$ and that

$$d(x, y) = \|x - y\|, \ x, y \in X$$

and that for all $\alpha \in [0, 1]$, $x, y \in X$,

$$(1 - \alpha)x \oplus \alpha y = (1 - \alpha)x + \alpha y.$$

Assume that $C \subset X$ is a non-empty convex set, m is a natural number, and that for $i = 1, \ldots, m$ and $T_i : C \to C$. Assume that

$$\widehat{\Delta} \in (0, , m^{-1}], \tag{5.1}$$

$M_0 > 0$ and the following assumption holds:

(A)   For each $\delta \in (0, 1)$ there exists $p_\delta \in B(0, M_0) \cap C$ such that for each $i \in \{1, \ldots, m\}$ and each $x \in C$,

$$\|T_i(x) - p_\delta\| \leq \|x - p_\delta\| + \delta.$$

For each $\delta \in (0, 1)$ let p_δ be as guaranteed by (A).
Now we describe our algorithm.
Initialization: Choose $x \in C \cap B(0, M_0)$, and set $x_0 = x$.
Iterative step: For each integer $i \geq 0$ choose

$$\alpha_i \in (0, 1)$$

$$\beta_{i,j} \geq \widehat{\Delta}, \ j = 1, \ldots, m$$

such that

$$\sum_{j=1}^{m} \beta_{i,j} = 1$$

and set

$$x_{i+1} = (1 - \alpha_i)x_i + \alpha_i \sum_{j=1}^{m} \beta_{i,j} T_j(x_i).$$

Proposition 5.1 *Assume that* $\{\Delta_i\}_{i=0}^{\infty} \subset [0, \infty)$, $\{\alpha_i\}_{i=0}^{\infty} \subset (0, 1)$,

$$\sum_{i=0}^{\infty} \Delta_i = \Delta < \infty, \tag{5.2}$$

$$x_0 \in B(0, M_0) \cap C, \tag{5.3}$$

$$\beta_{n,i} \geq 0, \ n = 0, 1, \ldots, \ i = 1, \ldots, m,$$

$$\sum_{i=1}^{m} \beta_{n,i} = 1, \ n = 0, 1, \ldots \tag{5.4}$$

and that for each integer $n \geq 0$,

$$x_{n+1} \in C \cap B((1 - \alpha_n)x_n + \alpha_n \sum_{i=1}^{n} \beta_{n,i} T_i(x_n), \Delta_n). \tag{5.5}$$

Then for each integer $n \geq 0$ and each $\delta \in (0, 1)$,

$$\|x_n\| \leq 3M_0 + \Delta, \ \|x_{n+1} - p_\delta\| \leq \Delta_n + \|x_n - p_\delta\| + \delta.$$

Proof Let $\delta \in (0, 1)$. Assumption (A), equations (5.3), and (5.5) and the convexity of the norm imply that for each integer $n \geq 0$,

$$\|x_{n+1} - p_\delta\| \leq \|x_{n+1} - (1 - \alpha_n)x_n - \sum_{i=1}^{m} \beta_{n,i} T_i(x_n)\|$$

$$+ \|(1 - \alpha_n)x_n + \alpha_n \sum_{i=1}^{m} \beta_{n,i} T_i(x_n) - p_\delta\|$$

$$\leq \Delta_n + (1 - \alpha_n)\|x_n - p_\delta\| + \alpha_n \|\sum_{i=1}^{m} \beta_{n,i} T_i(x_n) - p_\delta\|$$

$$\leq \Delta_n + (1 - \alpha_n)\|x_n - p_\delta\| + \alpha_n \sum_{i=1}^{m} \beta_{n,i} \|T_i(x_n) - p_\delta)\|$$

$$\leq \Delta_n + \|x_n - p_\delta\| + \delta.$$

Together with assumption (A) and equations (5.2), (5.3), this implies that for each integer $n \geq 0$,

$$\|x_n - p_\delta\| \leq \|x_0 - p_\delta\| + n\delta + \Delta \leq 2M_0 + n\delta + \Delta,$$

$$\|x_n\| \leq 3M_0 + n\delta + \Delta,$$

and since δ is any element of the interval $(0, 1)$, we have

$$\|x_n\| \leq 3M_0 + \Delta.$$

Proposition 5.1 is proved.

The following theorem shows that our algorithm generates approximate solution of the common fixed point problem under the presence of summable computational errors.

Theorem 5.2 *Assume that $\{\Delta_i\}_{i=0}^{\infty} \subset [0, \infty)$, $\{\alpha_i\}_{i=0}^{\infty} \subset (0, 1)$,*

$$\Delta = \sum_{i=0}^{\infty} \Delta_i < \infty, \tag{5.6}$$

$$\beta_{n,i} \in [\widehat{\Delta}, 1], \ n = 0, 1, \ldots, \ i = 1, \ldots, m,$$

$$\sum_{i=1}^{m} \beta_{n,i} = 1, \ n = 0, 1, \ldots, \tag{5.7}$$

$$x_0 \in B(0, M_0) \cap C, \tag{5.8}$$

for each integer $n \geq 0$,

$$x_{n+1} \in C \cap B((1 - \alpha_n)x_n + \alpha_n \sum_{i=1}^{n} \beta_{n,i} T_i(x_n), \Delta_n), \tag{5.9}$$

$\epsilon \in (0, 1)$, Q *is a natural number, and that*

$$\sum_{n=0}^{Q-1} \alpha_n (1 - \alpha_n)$$

$$> (8M_0 + 2\Delta)\widehat{\Delta}^{-1}\epsilon^{-1}\eta^{-1}(6M_0 + 2\Delta + 1, \epsilon(6M_0 + 1 + 2\Delta)^{-1}).$$

Then there exists an integer $n \in \{0, \ldots, Q\}$ *such that* $x_n \in F_\epsilon$.

Proof Proposition 5.1 and (5.6)–(5.9) imply that for each integer $n \geq 0$,

$$\|x_n\| \leq 3M_0 + \Delta. \tag{5.10}$$

Fix a positive number δ such that

$$\delta < 8^{-1}\epsilon\eta(6M_0 + 2\Delta + 1, 2^{-1}\epsilon(6M_0 + 1 + 2\Delta)^{-1})\alpha_i(1 - \alpha_i), \ i = 0, \ldots, Q. \tag{5.11}$$

We show that there exists $n \in \{0, \ldots, Q\}$ such that

$$x_n \in F_\epsilon.$$

Assume the contrary. Then for each integer $n \in \{0, \ldots, Q\}$,

$$x_n \notin F_\epsilon. \tag{5.12}$$

Let $n \in \{0, \ldots, Q\}$. In view of (5.12), there exists $j \in \{1, \ldots, m\}$ such that

$$\|x_n - T_j(x_n)\| > \epsilon. \tag{5.13}$$

In view of (5.7), (5.9), and the convexity of the norm,

$$\|x_{n+1} - p_\delta\| \le \|x_{n+1} - (1 - \alpha_n)x_n - \alpha_n \sum_{i=1}^{n} \beta_{n,i} T_i(x_n)\|$$

$$+ \|(1 - \alpha_n)x_n + \alpha_n \sum_{i=1}^{n} \beta_{n,i} T_i(x_n) - p_\delta\|$$

$$\le \Delta_n + \| \sum_{i=1}^{m} \beta_{n,i}((1 - \alpha_n)x_n + T_i(x_n) - p_\delta)\|$$

$$\le \Delta_n + \sum_{i=1}^{m} \beta_{n,i} \|(1 - \alpha_n)x_n + T_i(x_n) - p_\delta\|. \tag{5.14}$$

By assumption (B), for all $i = 1, \ldots, m$,

$$\|(1 - \alpha_n)x_n + \alpha_n T_i(x_n) - p_\delta\|$$

$$\le (1 - \alpha_n)\|x_n - p_\delta\| + \alpha_n \|T_i(x_n) - p_\delta\| \le \|x_n - p_\delta\| + \delta. \tag{5.15}$$

By assumption (B), equations (5.10), (5.11), (5.13), and Lemma 2.4 applied with $S = T_j, u = x_n, \alpha = \alpha_n, M = 3M_0 + \Delta, p = p_\delta, \gamma = \epsilon,$ and

$$v = (1 - \alpha_n)x_n + \alpha_n T_j(x_n),$$

we have

$$\|(1 - \alpha_n)x_n + \alpha_n T_j(x_n) - p_\delta\|$$

$$\le \|x_n - p_\delta\| - 4^{-1}\epsilon\alpha_n(1 - \alpha_n)\eta(6M_0 + 2\Delta + 1, \epsilon(6M_0 + 2\Delta + 1)^{-1}). \tag{5.16}$$

In view of (5.7) and (5.14)–(5.16),

$$\|x_{n+1} - p_\delta\| \le \Delta_n + \sum_{i=1}^{m} \beta_{n,i} \|(1 - \alpha_n)x_n + T_i(x_n) - p_\delta\|$$

$$\le \Delta_n + \beta_{n,j} \|(1 - \alpha_n)x_n + T_j(x_n) - p_\delta\|$$

$$+ \sum \{\beta_{n,i} \|(1 - \alpha_n)x_n + T_i(x_n) - p_\delta\| : i \in \{1, \ldots, m\} \setminus \{j\}\}$$

$$\le \Delta_n + \beta_{n,j}(\|x_n - p_\delta\| - 4^{-1}\epsilon\alpha_n(1 - \alpha_n)\eta(6M_0 + 2\Delta + 1, \epsilon(6M_0 + 2\Delta + 1)^{-1}))$$

$$+ \sum \{ \beta_{n,i}(\|x_n - p_\delta\| + \delta) : i \in \{1, \ldots, m\} \setminus \{j\}\}$$

$$\leq \|x_n - p_\delta\| - 4^{-1} \widehat{\Delta} \epsilon \alpha_n (1 - \alpha_n) \eta (6M_0 + 2\Delta + 1, \epsilon (6M_0 + 2\Delta + 1)^{-1}) + \Delta_n + \delta. \tag{5.17}$$

By assumption (A), (5.8) and (5.17),

$$2M_0 \geq \|x_0 - p_\delta\| \geq \|x_0 - p_\delta\| - \|x_{Q+1} - p_\delta\|$$

$$= \sum_{n=0}^{Q} (\|x_n - p_\delta\| - \|x_{n+1} - p_\delta\|)$$

$$\geq \sum_{n=0}^{Q} \alpha_n (1 - \alpha_n) 4^{-1} \widehat{\Delta} \epsilon \eta (6M_0 + 2\Delta + 1, \epsilon_0 (6M_0 + 2\Delta + 1)^{-1})$$

$$- \sum_{n=0}^{Q} \Delta_n - \delta Q.$$

Since δ is any positive sufficient small number, the relation above implies that

$$\sum_{n=0}^{Q} \alpha_n (1 - \alpha_n) \leq (8M_0 + 2\Delta) \widehat{\Delta}^{-1} \epsilon^{-1} \eta^{-1} (6M_0 + 2\Delta + 1, \epsilon (6M_0 + 2\Delta + 1)^{-1}).$$

This contradicts our definition of Q. The contradiction we have reached proves Theorem 5.2.

Theorem 5.2 implies the following result.

Theorem 5.3 *Assume that* $\{\Delta_i\}_{i=0}^{\infty} \subset [0, \infty)$,

$$\Delta = \sum_{i=0}^{\infty} \Delta_i < \infty,$$

$\{\alpha_i\}_{i=0}^{\infty} \subset (0, 1)$,

$$\beta_{n,i} \in [\widehat{\Delta}, 1], \; n = 0, 1, \ldots, \; i = 1, \ldots, m,$$

$$\sum_{i=1}^{m} \beta_{n,i} = 1, \; n = 0, 1, \ldots,$$

$$x_0 \in B(0, M_0) \cap C,$$

for each integer $n \geq 0$,

$$x_{n+1} \in C \cap B((1 - \alpha_n)x_n + \alpha_n \sum_{i=1}^{n} \beta_{n,i} T_i(x_n), \Delta_n),$$

$$\sum_{n=0}^{\infty} \alpha_n(1 - \alpha_n) = \infty.$$

Then there exists a strictly increasing sequence of natural numbers n_k *such that for each integer* $k \geq 1$, $x_{n_k} \in F_{1/k}$.

5.2 Exact Iterates

The results of this section describe the behavior of exact iterates of our algorithm.

Theorem 5.4 *Assume that* $\Lambda \in (0, 2^{-1})$,

$$\{\alpha_i\}_{i=0}^{\infty} \subset (\Lambda, 1 - \Lambda), \tag{5.18}$$

$$\beta_{n,i} \in [\widehat{\Delta}, 1], \ n = 0, 1, \ldots, \ i = 1, \ldots, m, \tag{5.19}$$

$$\sum_{i=1}^{m} \beta_{n,i} = 1, \ n = 0, 1, \ldots, \tag{5.20}$$

$$x_0 \in B(0, M_0) \cap C, \tag{5.21}$$

for each integer $n \geq 0$,

$$x_{n+1} = (1 - \alpha_n)x_n + \alpha_n \sum_{i=1}^{n} \beta_{n,i} T_i(x_n) \tag{5.22}$$

and $\epsilon \in (0, 1)$. *Then*

$$Card(\{n \in \{0, 1, \ldots\} : x_n \notin \tilde{F}_\epsilon\})$$

$$\leq 8M_0 \widehat{\Delta}^{-1} \epsilon^{-1} \eta^{-1}(6M_0 + 1, \epsilon(6M_0 + 1)^{-1}).$$

Proof Proposition 5.1 and (5.18)–(5.22) imply that for each integer $n \geq 0$,

$$\|x_n\| \leq 3M_0. \tag{5.23}$$

Fix a positive number δ such that

$$\delta < 8^{-1}\epsilon\Lambda^2\eta(6M_0 + 1, \epsilon(6M_0 + 1)^{-1}), \tag{5.24}$$

$$E = \{n \in \{0, 1, \ldots\} : x_n \notin F_\epsilon\}. \tag{5.25}$$

Let Q be a natural number. Assume that $n \in E$. In view of (5.25), $x_n \notin F_\epsilon$, and there exists $j_n \in \{1, \ldots, m\}$ such that

$$\|x_n - T_{j_n}(x_n)\| > \epsilon. \tag{5.26}$$

In view of (5.20), (5.22), and the convexity of the norm,

$$\|x_{n+1} - p_\delta\| = \|(1 - \alpha_n)x_n + \alpha_n \sum_{i=1}^{n} \beta_{n,i} T_i(x_n) - p_\delta\|$$

$$= \|(1 - \alpha_n) \sum_{i=1}^{m} \beta_{n,i} x_n + \alpha_n \sum_{i=1}^{m} \beta_{n,i} T_i(x_n) - p_\delta\|$$

$$= \|\sum_{i=1}^{m} \beta_{n,i}((1 - \alpha_n)x_n + \alpha_n T_i(x_n) - p_\delta)\|$$

$$\leq \sum_{i=1}^{m} \beta_{n,i}\|(1 - \alpha_n)x_n + \alpha_n T_i(x_n) - p_\delta\|. \tag{5.27}$$

By assumption (A), equations (5.18), (5.23), (5.26), and Lemma 2.4 applied with $S = T_{j_n}$, $u = x_n$, $\alpha = \alpha_n$, $M = 3M_0$, $p = p_\delta$, $\gamma = \epsilon$, and $v = (1 - \alpha_n)x_n + \alpha_n T_{j_n}(x_n)$, we have

$$\|(1 - \alpha_n)x_n + \alpha_n T_{j_n}(x_n) - p_\delta\|$$

$$\leq \|x_n - p_\delta\| - 4^{-1}\epsilon\alpha_n(1 - \alpha_n)\eta(6M_0 + 1, \epsilon(6M_0 + 1)^{-1}). \tag{5.28}$$

In view of assumption (B), (5.19), (5.20), (5.27), (5.28),

$$\|x_{n+1} - p_\delta\| \leq \sum_{i=1}^{m} \beta_{n,i}\|(1 - \alpha_n)x_n + \alpha_n T_i(x_n) - p_\delta\|$$

$$\leq \|x_n - p_\delta\| + \sum_{i=1}^{m} \beta_{n,i}(\|(1 - \alpha_n)x_n + \alpha_n T_i(x_n) - p_\delta\| - \|x_n - p_\delta\|)$$

$$\leq \|x_n - p_\delta\| + \beta_{n,j_n}(\|(1-\alpha_n)x_n + \alpha_n T_{j_n}(x_n) - p_\delta\| - \|x_n - p_\delta\|)$$

$$+ \sum\{\beta_{n,i}(\|(1-\alpha_n)x_n + \alpha_n T_i(x_n) - p_\delta\| - \|x_n - p_\delta\|) : i \in \{1, \ldots, m\} \setminus \{j_n\}\}$$

$$\leq \|x_n - p_\delta\| - 4^{-1}\widehat{\Delta}\epsilon\alpha_n(1-\alpha_n)\eta(6M_0 + 1, \epsilon(6M_0 + 1)^{-1}) + \delta. \qquad (5.29)$$

By assumption (A), Proposition 5.1, (5.18), (5.21), and (5.29),

$$2M_0 \geq \|x_0 - p_\delta\| \geq \|x_0 - p_\delta\| - \|x_{Q+1} - p_\delta\|$$

$$= \sum_{n=0}^{Q}(\|x_n - p_\delta\| - \|x_{n+1} - p_\delta\|)$$

$$= \sum\{\|x_n - p_\delta\| - \|x_{n+1} - p_\delta\| : n \in E \cap [0, Q]\}$$

$$+ \sum\{\|x_n - p_\delta\| - \|x_{n+1} - p_\delta\| : n \in \{0, \ldots, Q\} \setminus E\}$$

$$\geq 4^{-1}\widehat{\Delta}\epsilon\Lambda^2\eta(6M_0 + 1, \epsilon(6M_0 + 1)^{-1})\mathrm{Card}(E \cap [0, Q]) - \delta(Q + 1).$$

Since δ is any positive sufficient small number, the relation above implies that

$$\mathrm{Card}(E \cap [0, Q]) \leq 8M_0\widehat{\Delta}^{-1}\epsilon^{-1}\Lambda^{-2}\eta^{-1}(6M_0 + 1, \epsilon(6M_0 + 1)^{-1}).$$

Since Q is any natural number, we conclude that

$$\mathrm{Card}(E) \leq 8M_0\widehat{\Delta}^{-1}\epsilon^{-1}\Lambda^{-2}\eta^{-1}(6M_0 + 1, \epsilon(6M_0 + 1)^{-1}).$$

Theorem 5.4 is proved.

Theorem 5.5 *Assume that* $\Lambda \in (0, 2^{-1})$,

$$\|T_i(x) - T_i(y)\| \leq \|x - y\|, \ x, y \in C, \ i = 1, \ldots, m, \qquad (5.30)$$

$$\beta_i \in [\widehat{\Delta}, 1], \ i = 1, \ldots, m, \ \sum_{i=1}^{m}\beta_i = 1, \qquad (5.31)$$

$$\alpha \in (\Lambda, 1 - \Lambda), \qquad (5.32)$$

$$x_0 \in B(0, M_0) \cap C, \qquad (5.33)$$

for each integer n ≥ 0,

$$x_{n+1} = (1 - \alpha)x_n + \alpha \sum_{i=1}^{m} \beta_i T_i(x_n) \tag{5.34}$$

and $\epsilon \in (0, 1)$. Let

$$0 < \epsilon_0 < 4^{-1}\epsilon\widehat{\Delta}\Lambda^2\eta(6M_0 + 1, \epsilon(6M_0 + 1)^{-1}), \tag{5.35}$$

$$N_0 = \lfloor 8M_0\epsilon_0^{-1}\Lambda^{-2}\eta^{-1}(6M_0 + 1, \epsilon_0(6M_0 + 1)^{-1})\widehat{\Delta}^{-1}\rfloor + 1. \tag{5.36}$$

Then for each integer $i \geq N_0$, $x_i \in F_\epsilon$.

Proof Set

$$S_0(y) = (1 - \alpha)y + \alpha \sum_{i=1}^{m} \beta_i T_i(y). \tag{5.37}$$

It follows from (5.30), (5.31), (5.37), and the convexity of the norm that for each $u, v \in C$,

$$\|S_0(u) - S_0(v)\|$$

$$= \left\|(1 - \alpha)u + \alpha \sum_{i=1}^{m} \beta_i T_i(u) - (1 - \alpha)v - \alpha \sum_{i=1}^{m} \beta_i T_i(v)\right\|$$

$$\leq (1 - \alpha)\|u - v\| + \alpha \left\|\sum_{i=1}^{m} \beta_i T_i(u) - \sum_{i=1}^{m} \beta_i T_i(v)\right\|$$

$$\leq (1 - \alpha)\|u - v\| + \alpha \sum_{i=1}^{m} \beta_i \|T_i(u) - T_i(v)\| \leq \|u - v\|. \tag{5.38}$$

By (5.34) and (5.37), for each integer $n \geq 0$,

$$x_{n+1} = S_0(x_n). \tag{5.39}$$

Proposition 5.1 implies that

$$\|x_k\| \leq 3M_0, \quad k = 0, 1, \ldots. \tag{5.40}$$

Set

$$E = \{n \in \{0, 1, \ldots\} : x_n \notin F_{\epsilon_0}\}. \tag{5.41}$$

Theorem 5.4 and and equations (5.31)–(5.34), (5.36), (5.41) imply that

$$\mathrm{Card}(E) \leq 8M_0\epsilon_0^{-1}\Lambda^{-2}\eta^{-1}(6M_0 + 1, \epsilon_0(6M_0 + 1)^{-1})\widehat{\Delta}^{-1}$$

and

$$\mathrm{Card}(E) < N_0.$$

Thus there exists an integer

$$n_0 \in [1, N_0] \tag{5.42}$$

such that

$$x_{n_0} \in F_{\epsilon_0}, \ \|x_{n_0} - T_i(x_{n_0})\| \leq \epsilon_0, \ i = 1, \ldots, m. \tag{5.43}$$

By (5.31), (5.37), (5.43), and the convexity of the norm,

$$\|x_{n_0} - S_0(x_{n_0})\| = \|x_{n_0} - (1 - \alpha)x_{n_0} - \alpha \sum_{i=1}^{m} \beta_i T_i(x_{n_0})\|$$

$$\leq \alpha\|x_{n_0} - \sum_{i=1}^{m} \beta_i T_i(x_{n_0})\| \leq \alpha \sum_{i=1}^{m} \beta_i \|x_{n_0} - T_i(x_{n_0})\| \leq \alpha\epsilon_0 \leq \epsilon_0. \tag{5.44}$$

In view of (5.38), (5.39), and (5.44), for each integer $n \geq 0$,

$$\|x_{n+1} - S_0(x_{n+1})\| = \|S_0(x_n) - S_0(S_0(x_n))\| \leq \|x_n - S_0(x_n)\|$$

and for each integer $n \geq n_0$,

$$\|x_{n+1} - x_n\| = \|x_n - S_0(x_n)\| \leq \|x_{n_0} - S_0(x_{n_0})\| \leq \epsilon_0. \tag{5.45}$$

Let $n \geq n_0$ be an integer. We show $x_n \in F_\epsilon$. Assume the contrary. Then there exists $j \in \{1, \ldots, m\}$ such that

$$\|x_n - T_j(x_n)\| > \epsilon. \tag{5.46}$$

Fix a positive number

$$\delta < 4^{-1}\epsilon_1\Lambda^2\eta(6M_0 + 1, \epsilon(6M_0 + 1)^{-1}). \tag{5.47}$$

Assumption (A) implies that for each $i \in \{1, \dots, m\}$,

$$\|T_i(x_n) - p_\delta\| \le \|p_\delta - x_n\| + \delta,$$

$$\|(1 - \alpha)x_n + \alpha T_i(x_n) - p_\delta\| \le \|x_n - p_\delta\| + \delta. \tag{5.48}$$

By assumption (A), equations (5.32), (5.40), (5.46), (5.47), and Lemma 2.4 applied with $S = T_j, u = x_n, M = 3M_0, p = p_\delta, \gamma = \epsilon$, and

$$v = (1 - \alpha_n)x_n + \alpha_n T_j(x_n),$$

we have

$$\|(1 - \alpha_n)x_n + \alpha_n T_j(x_n) - p_\delta\| \le \|x_n - p_\delta\| - 4^{-1}\epsilon \Lambda^2 \eta(6M_0 + 1, \epsilon(6M_0 + 1)^{-1}). \tag{5.49}$$

It follows from (5.31), (5.34), (5.48), (5.49), and the convexity of the norm that

$$\|x_{n+1} - p_\delta\| - \|x_n - p_\delta\|$$

$$\le \|(1 - \alpha)x_n + \alpha \sum_{i=1}^{m} \beta_i T_i(x_n) - p_\delta)\| - \|x_n - p_\delta\|$$

$$= \|\sum_{i=1}^{m} \beta_i((1 - \alpha)x_n + \alpha T_i(x_n) - p_\delta)\| - \|x_n - p_\delta\|$$

$$\le \sum_{i=1}^{m} \beta_i \|(1 - \alpha)x_n + \alpha T_i(x_n) - p_\delta\| - \|x_n - p_\delta\|$$

$$\le \beta_j(\|(1 - \alpha)x_n + \alpha T_j(x_n) - p_\delta)\| - \|x_n - p_\delta\|)$$

$$+ \sum \{\beta_i(\|(1 - \alpha)x_n + \alpha T_i(x_n) - p_\delta\| - \|x_n - p_\delta\|) : i \in \{1, \dots, m\} \setminus \{j\}\}$$

$$\le -4^{-1}\widehat{\Delta}\epsilon \Lambda^2 \eta(6M_0 + 1, \epsilon(6M_0 + 1)^{-1}) + \delta. \tag{5.50}$$

In view of (5.45) and (5.50),

$$\epsilon_0 \ge \|x_{n+1} - x_n\| \ge \|x_n - p_\delta\| - \|x_{n+1} - p_\delta\|$$

$$\ge 4^{-1}\epsilon \widehat{\Delta}\Lambda^2 \eta(6M_0 + 1, \epsilon(6M_0 + 1)^{-1}) - \delta.$$

Since δ is any positive sufficiently small number, we conclude that

$$\epsilon_0 \geq 4^{-1}\epsilon\widehat{\Delta}\Lambda^2\eta(6M_0+1,\epsilon(6M_0+1)^{-1}).$$

This contradicts (5.35). The contradiction we have reached proves that $x_n \in F_\epsilon$ and completes the proof of Theorem 5.5.

5.3 Inexact Iterates with Summable Errors

In this section we prove two theorems which describe the behavior of inexact iterates of our algorithm with summable errors.

Theorem 5.6 *Assume that $\Lambda \in (0, 2^{-1})$,*

$$\{\alpha_i\}_{i=0}^\infty \subset (\Lambda, 1-\Lambda), \tag{5.51}$$

$$\beta_{n,i} \in [\widehat{\Delta}, 1], \ n = 0, 1, \ldots, \ i = 1, \ldots, m, \tag{5.52}$$

$$\sum_{i=1}^m \beta_{n,i} = 1, \ n = 0, 1, \ldots, \tag{5.53}$$

$\{\Delta_i\}_{i=0}^\infty \subset [0, \infty),$

$$\Delta = \sum_{i=0}^\infty \Delta_i < \infty, \tag{5.54}$$

$$x_0 \in B(0, M_0) \cap C, \tag{5.55}$$

for each integer $n \geq 0$,

$$x_{n+1} \in C \cap B\left((1-\alpha_n)x_n + \alpha_n \sum_{i=1}^n \beta_{n,i}T_i(x_n), \Delta_n\right), \tag{5.56}$$

and $\epsilon \in (0, 1)$. Then

$$Card(\{n \in \{0, 1, \ldots\} : \ x_n \notin \tilde{F}_\epsilon\})$$

$$\leq (8M_0 + 4\Delta)\widehat{\Delta}^{-1}\epsilon^{-1}\Lambda^{-2}\eta^{-1}(6M_0+2\Delta+1, \epsilon(6M_0+2\Delta+1)^{-1}).$$

Proof Proposition 5.1 and (5.52), (5.56) imply that for each integer $n \geq 0$,

$$\|x_n\| \leq 3M_0 + \Delta. \tag{5.57}$$

Fix a positive number δ such that

$$\delta < 8^{-1}\epsilon\Lambda^2\eta(6M_0 + 1 + 2\Delta, \epsilon(6M_0 + 1 + 2)\Delta)^{-1}) \tag{5.58}$$

and set

$$E = \{n \in \{0, 1, \dots\} : x_n \notin F_\epsilon\}. \tag{5.59}$$

Assume that $n \in E$. In view of (5.59), $x_n \notin F_\epsilon$, and there exists $j \in \{1, \dots, m\}$ such that

$$\|x_n - T_j(x_n)\| > \epsilon. \tag{5.60}$$

In view of (5.53), (5.56), and the convexity of the norm,

$$\|x_{n+1} - p_\delta\| \leq \|x_{n+1} - (1 - \alpha_n)x_n - \alpha_n \sum_{i=1}^{m} \beta_{n,i} T_i(x_n)\|$$

$$+ \|(1 - \alpha_n)x_n + \alpha_n \sum_{i=1}^{n} \beta_{n,i} T_i(x_n) - p_\delta\|$$

$$\leq \Delta_n + \|\sum_{i=1}^{m} \beta_{n,i}((1 - \alpha_n)x_n + \alpha_n T_i(x_n) - p_\delta)\|$$

$$\leq \Delta_n + \sum_{i=1}^{m} \beta_{n,i}\|(1 - \alpha_n)x_n + \alpha_n T_i(x_n) - p_\delta)\|. \tag{5.61}$$

By assumption (A), for every $i \in \{1, \dots, m\}$,

$$\|(1 - \alpha_n)x_n + \alpha_n T_i(x_n) - p_\delta\|$$

$$\leq (1 - \alpha_n)\|x_n - p_\delta\| + \alpha_n\|T_i(x_n) - p_\delta\| \leq \|x_n - p_\delta\| + \delta. \tag{5.62}$$

By assumption (A), equations (5.51), (5.57), (5.58), (5.60) and Lemma 2.4 applied with $S = T_j$, $u = x_n$, $\alpha = \alpha_n$, $M = 3M_0 + \Delta$, $p = p_\delta$, $\gamma = \epsilon$, and $v = (1 - \alpha_n)x_n + \alpha_n T_j(x_n)$ we have

$$\|(1 - \alpha_n)x_n + \alpha_n T_j(x_n) - p_\delta\|$$

$$\leq \|x_n - p_\delta\| - 4^{-1}\epsilon\alpha_n(1 - \alpha_n)\eta(6M_0 + 2\Delta + 1, \epsilon(6M_0 + 2\Delta + 1)^{-1}). \tag{5.63}$$

In view (5.51), (5.52), (5.57), (5.61)–(5.63),

$$\|x_{n+1} - p_\delta\| \le \Delta_n + \sum_{i=1}^{m} \beta_{n,i} \|(1-\alpha_n)x_n + \alpha_n T_i(x_n) - p_\delta\|$$

$$\le \Delta_n + \|x_n - p_\delta\| + \sum_{i=1}^{m} \beta_{n,i}(\|(1-\alpha_n)x_n + \alpha_n T_i(x_n) - p_\delta\| - \|x_n - p_\delta\|)$$

$$\le \Delta_n + \|x_n - p_\delta\| + \beta_{n,j}(\|(1-\alpha_n)x_n + \alpha_n T_j(x_n) - p_\delta\| - \|x_n - p_\delta\|)$$

$$+ \sum\{\beta_{n,i}(\|(1-\alpha_n)x_n + \alpha_n T_i(x_n) - p_\delta\| - \|x_n - p_\delta\|) : i \in \{1,\dots,m\} \setminus \{j\}\}$$

$$\le \Delta_n + \|x_n - p_\delta\| - 4^{-1}\widehat{\Delta}\epsilon\Lambda^2\eta(6M_0 + 2\Delta + 1, \epsilon(6M_0 + 2\Delta + 1)^{-1}) - \delta. \tag{5.64}$$

Thus (5.64) holds for each $n \in E$.

Let Q be a natural number. By assumption (A), (5.55), (5.64), and Proposition 5.1,

$$2M_0 \ge \|x_0 - p_\delta\| \ge \|x_0 - p_\delta\| - \|x_{Q+1} - p_\delta\|$$

$$= \sum_{n=0}^{Q} (\|x_n - p_\delta\| - \|x_{n+1} - p_\delta\|)$$

$$= \sum\{\|x_n - p_\delta\| - \|x_{n+1} - p_\delta\| : n \in E \cap [0, Q]\}$$

$$+ \sum\{\|x_n - p_\delta\| - \|x_{n+1} - p_\delta\| : n \in \{0,\dots,Q\} \setminus E\}$$

$$\ge 4^{-1}\widehat{\Delta}\epsilon\Lambda^2\eta(6M_0 + 2\Delta + 1, \epsilon(6M_0 + 2\Delta + 1)^{-1})\text{Card}(E \cap [0, Q]) - \sum_{n=0}^{Q}\Delta_n - \delta Q.$$

Since δ is any positive sufficient small number, the relation above implies that

$$\text{Card}(E \cap [0, Q])$$

$$\le 4(2M_0 + \Delta)\widehat{\Delta}^{-1}\epsilon^{-1}\Lambda^{-2}\eta^{-1}(6M_0 + 1 + 2\Delta, \epsilon(6M_0 + 2\Delta + 1)^{-1}).$$

Since Q is any natural number, we conclude that

$$\text{Card}(E) \le (8M_0 + 4\Delta)\widehat{\Delta}^{-1}\epsilon^{-1}\Lambda^{-2}\eta^{-1}(6M_0 + 1 + 2\Delta, \epsilon(6M_0 + 1 + 2\Delta)^{-1}).$$

Theorem 5.6 is proved.

Theorem 5.7 *Assume that* $\Lambda \in (0, 2^{-1})$, $\{\Delta_i\}_{i=0}^{\infty} \subset [0, \infty)$,

$$\Delta = \sum_{i=0}^{\infty} \Delta_i < \infty, \tag{5.65}$$

$$\|T_i(x) - T_i(y)\| \le \|x - y\|, \ x, y \in C, \ i = 1, \dots, m, \tag{5.66}$$

$$\alpha \in (\Lambda, 1 - \Lambda), \tag{5.67}$$

$$\beta_i \in [\widehat{\Delta}, 1], \ i = 1, \dots, m, \ \sum_{i=1}^{m} \beta_i = 1, \tag{5.68}$$

$$x_0 \in B(0, M_0) \cap C, \tag{5.69}$$

for each integer $n \ge 0$,

$$x_{n+1} \in B\left((1 - \alpha)x_n + \alpha \sum_{i=1}^{m} \beta_i T_i(x_n), \Delta_n\right) \cap C, \tag{5.70}$$

$\epsilon \in (0, 1)$, n_0 *is a natural number,*

$$\sum_{i=n_0}^{\infty} \Delta_i \le \epsilon/4, \tag{5.71}$$

$$0 < \epsilon_0 < 16^{-1} \epsilon \widehat{\Delta} \Lambda^2 \eta (18M_0 + 6\Delta + 1, 4^{-1} \epsilon (18M_0 + 6\Delta + 1)^{-1}),$$

and

$$N_0 = \lfloor (24M_0 + 8\Delta) \widehat{\Delta}^{-1} \epsilon_0^{-1} \Lambda^{-2} \eta^{-1} (18M_0 + 6\Delta + 1, \epsilon_0(18M_0 + 6\Delta + 1)^{-1}) \rfloor + 1.$$

Then for each integer $i \ge n_0 + N_0$, $x_i \in F_\epsilon$.

Proof Proposition 5.1 and equations (5.65) and (5.67)–(5.70)) imply that for each integer $n \ge 0$,

$$\|x_n\| \le 3M_0 + \Delta. \tag{5.72}$$

Define a sequence $\{y_i\}_{i=n_0}^{\infty} \subset C$ such that

$$y_{n_0} = x_{n_0} \tag{5.73}$$

and that for each integer $n \geq n_0$,

$$y_{n+1} = (1 - \alpha)y_n + \alpha \sum_{i=1}^{m} \beta_i T_i(y_n). \tag{5.74}$$

By Theorem 5.5, equations (5.66), (5.67), (5.72), (5.74) and the choice of ϵ_0, N_0, for each integer $n \geq n_0 + N_0$,

$$y_n \in F_{\epsilon/4}. \tag{5.75}$$

We show that for each integer $n \geq n_0 + 1$,

$$\|x_n - y_n\| \leq \sum_{i=n_0}^{n-1} \Delta_i. \tag{5.76}$$

By (5.70), (5.73), and (5.74),

$$\|x_{n_0+1} - y_{n_0+1}\| \leq \Delta_{n_0}.$$

Assume that $k \geq n_0 + 1$ is an integer and (5.76) holds for $n = k$. Equations (5.66), (5.68), (5.70), and (5.76) imply that

$$\|x_{k+1} - y_{k+1}\| \leq \|x_{k+1} - (1 - \alpha)x_k - \alpha \sum_{i=1}^{m} \beta_i T_i(x_k)\|$$

$$+ \|(1 - \alpha)x_k + \alpha \sum_{i=1}^{m} \beta_i T_i(x_k) - (1 - \alpha)y_k - \alpha \sum_{i=1}^{m} \beta_i T_i(y_k)\|$$

$$\leq \Delta_k + (1 - \alpha)\|x_k - y_k\| + \alpha \sum_{i=1}^{m} \beta_i \|x_k - y_k\|$$

$$\leq \Delta_k + \|x_k - y_k\| \leq \sum_{i=n_0}^{k} \Delta_i,$$

and (5.76) holds for $k + 1$ too. Thus by induction we showed that (5.76) holds for each integer $n \geq n_0$. By (5.71) and (5.76), for each integer $n \geq n_0$,

$$\|x_n - y_n\| \leq \epsilon/4. \tag{5.77}$$

Assume that $n \geq n_0 + N_0$ is an integer and $s \in \{1, \ldots, m\}$. In view of (5.66), (5.75), and (5.77),

$$\|x_n - T_s(x_n)\| \leq \|x_n - y_n\| + \|y_n - T_s(y_n)\| + \|T_s(y_n) - T_s(x_n)\|$$

$$\leq 2\|x_n - y_n\| + \epsilon/4 < \epsilon$$

and $x_n \in F_\epsilon$. Theorem 5.7 is proved.

5.4   Inexact Iterates with Nonsummable Errors

In this section we prove two theorems which describe the behavior of inexact iterates of our algorithm with nonsummable errors. These results show that if computational errors are small enough, then our method generates approximate solution belonging to the set $\tilde{F}_\gamma$ with $\gamma > 0$. Our first result shows the dependence of γ on our computational errors and calculates the number of iterates which should be done in order to obtain this approximate solution.

Theorem 5.8 *Let* $\Lambda \in (0, 2^{-1})$, $\epsilon_0 \in (0, 1)$,

$$0 < \delta_0 < 16^{-1}\epsilon_0\Lambda^2\widehat{\Delta}\eta(6M_0 + 3, \epsilon_0(6M_0 + 3)^{-1}), \tag{5.78}$$

$$n_0 = \lfloor 8(2M_0 + 1)\widehat{\Delta}^{-1}\epsilon_0^{-1}\Lambda^{-2}\eta^{-1}(6M_0 + 3, \epsilon_0(6M_0 + 3)^{-1})\rfloor + 1. \tag{5.79}$$

Assume that

$$\{\alpha_i\}_{i=0}^{\infty} \subset [\Lambda, 1 - \Lambda], \tag{5.80}$$

$$\beta_{n,i} \in [\widehat{\Delta}, 1], \ n = 0, 1, \ldots, \ i = 1, \ldots, m, \tag{5.81}$$

$$\sum_{i=1}^{m} \beta_{n,i} = 1, \ n = 0, 1, \ldots, \tag{5.82}$$

$$x_0 \in B(0, M_0) \cap C \tag{5.83}$$

and that for each integer $n \geq 0$,

$$x_{n+1} \in C \cap B((1 - \alpha_n)x_n + \alpha_n \sum_{i=1}^{n} \beta_{n,i}T_i(x_n), \delta_0). \tag{5.84}$$

Then there exists an integer $q \in [1, n_0]$ such that

$$\|x_i\| \leq 3M_0 + 1, \quad i = 0, \ldots, q$$

and $x_q \in F_{\epsilon_0}$.

Proof Fix

$$\delta \in (0, 2^{-1}\delta_0). \tag{5.85}$$

In view of assumption (A), equations (5.78), (5.82)–(5.85), and the convexity of the norm,

$$\|x_1 - p_\delta\| = \|x_1 - (1 - \alpha_0)x_0 - \alpha_0 \sum_{i=1}^{m} \beta_{0,i} T_i(x_0)\|$$

$$+ \|(1 - \alpha_0)x_0 + \alpha_0 \sum_{i=1}^{m} \beta_{0,i} T_i(x_0) - p_\delta\|$$

$$\leq \delta_0 + (1 - \alpha_0)\|x_0 - p_\delta\| + \alpha_0 \sum_{i=1}^{m} \beta_{0,i} T_i(x_0) - p_\delta\|$$

$$\leq \|x_0 - p_\delta\| + \delta_0 + \delta \leq 2M_0 + 1. \tag{5.86}$$

Assume that s is a natural number and that for each integer $k \in [1, s]$,

$$x_k \notin F_{\epsilon_0}. \tag{5.87}$$

Assume that $k \in \{1, \ldots, s\}$ as an integer and that

$$\|x_k - p_\delta\| \leq 2M_0 + 1. \tag{5.88}$$

(In view of (5.86), inequality (5.88) holds for $k = 1$.) Assumption (A) implies that

$$\|x_k\| \leq 3M_0 + 1. \tag{5.89}$$

By (5.87), there exists $j \in \{1, \ldots, m\}$ such that

$$\|x_k - T_j(x_k)\| > \epsilon_0. \tag{5.90}$$

By assumption (A), equations (5.78), (5.80), (5.85), (5.89), (5.90) and Lemma 2.4 applied with $S = T_j$, $u = x_k$, $\alpha = \alpha_k$, $M = 3M_0 + 1$, $p = p_\delta$, $\gamma = \epsilon_0$, and

$$v = (1 - \alpha_k)x_k + \alpha_k T_j(x_k),$$

we have

$$\|(1 - \alpha_k)x_k + \alpha_k T_j(x_k) - p_\delta\|$$

$$\leq \|x_k - p_\delta\| - 4^{-1}\epsilon_0 \Lambda^2 \eta(6M_0 + 3, \epsilon_0(6M_0 + 3)^{-1}). \tag{5.91}$$

It follows from assumption (A) that for each $s \in \{1, \ldots, m\}$,

$$\|(1 - \alpha_k)x_k + \alpha_k T_i(x_k) - p_\delta\|$$

$$\leq (1 - \alpha_k)\|x_k - p_\delta\| + \alpha_k\|T_i(x_k) - p_\delta\| \leq \|x_k - p_\delta\| + \delta. \tag{5.92}$$

Equations (5.78), (5.82), (5.84), (5.85), (5.91), (5.92) and the convexity of the norm imply that

$$\|x_{k+1} - p_\delta\| \leq \|x_{k+1} - (1 - \alpha_k)x_k - \alpha_k \sum_{i=1}^{m} \beta_{k,i} T_i(x_k)\|$$

$$+ \|(1 - \alpha_k)x_k + \alpha_k \sum_{i=1}^{m} \beta_{k,i} T_i(x_k) - p_\delta\|$$

$$\leq \delta_0 + \|\sum_{i=1}^{m} \beta_{k,i}((1 - \alpha_k)x_k + \alpha_k T_i(x_k) - p_\delta)\|$$

$$\leq \delta_0 + \sum_{i=1}^{m} \beta_{k,i}\|(1 - \alpha_k)x_k + \alpha_k T_i(x_k) - p_\delta\|$$

$$\leq \delta_0 + \beta_{k,j}\|(1 - \alpha_k)x_k + \alpha_k T_j(x_k) - p_\delta\|$$

$$+ \sum\{\beta_{k,i}\|(1 - \alpha_k)x_k + \alpha_k T_i(x_k) - p_\delta\| : i \in \{1, \ldots, m\} \setminus \{j\}\}$$

$$\leq \delta_0 + \beta_{k,j}(\|x_k - p_\delta\| - 4^{-1}\epsilon_0 \Lambda^2 \eta(6M_0 + 3, \epsilon_0(6M_0 + 3)^{-1}))$$

$$+ \sum\{\beta_{k,i}(\|x_k - p_\delta\| + \delta) : i \in \{1, \ldots, m\} \setminus \{j\}\}$$

$$\leq \delta_0 + \|x_k - p_\delta\| + \delta - 4^{-1}\beta_{k,j}\epsilon_0 \Lambda^2 \eta(6M_0 + 3, \epsilon_0(6M_0 + 3)^{-1})$$

$$\leq \|x_k - p_\delta\| + 2\delta_0 - 4^{-1}\widehat{\Delta}\epsilon_0 \Lambda^2 \eta(6M_0 + 3, \epsilon_0(6M_0 + 3)^{-1})$$

$$\leq \|x_k - p_\delta\| - 8^{-1}\widehat{\Delta}\epsilon_0 \Lambda^2 \eta(6M_0 + 3, \epsilon_0(6M_0 + 3)^{-1}).$$

Thus by induction we showed that for each $k \in \{1, \ldots, s+1\}$

$$\|x_{k+1} - p_\delta\| \leq 2M_0 + 1$$

and for each $k \in \{1, \ldots, s\}$,

$$\|x_{k+1} - p_\delta\| \leq \|x_k - p_\delta\| - 8^{-1}\epsilon_0 \widehat{\Delta}\Lambda^2 \eta(6M_0 + 3, \epsilon_0(6M_0 + 3)^{-1}). \qquad (5.93)$$

In view of (5.86) and (5.93),

$$8^{-1}\epsilon_0 \widehat{\Delta}\Lambda^2 \eta(6M_0 + 3, \epsilon_0(6M_0 + 3)^{-1})s \leq \sum_{k=1}^{s}(\|x_k - p_\delta\| - \|x_{k+1} - p_\delta\|)$$

$$\leq \|x_1 - p_\delta\| \leq 2M_0 + 1,$$

$$s \leq 8(2M_0 + 1)\epsilon_0^{-1}\widehat{\Delta}^{-1}\Lambda^{-2}\eta^{-1}(6M_0 + 3, \epsilon_0(6M_0 + 3)^{-1}). \qquad (5.94)$$

Thus we have shown that the following property holds:

(P) If s is a natural number and for each $k \in \{1, \ldots, s\}$, $x_k \notin F_{\epsilon_0}$, then (5.94) holds and

$$\|x_j - p_\delta\| \leq 2M_0 + 1, \quad j = 1, \ldots, s+1.$$

This implies that there exists an integer $q \in [1, n_0]$ such that

$$x_q \in F_\epsilon,$$

for each integer k satisfying $1 \leq k < q$, $x_k \notin F_\epsilon$, for each integer $k \in \{1, \ldots, q\}$,

$$\|x_k - p_\delta\| \leq 2M_0 + 1, \quad \|x_k\| \leq 3M_0 + 1.$$

Theorem 5.8 is proved.

Theorem 5.8 implies the following result.

Theorem 5.9 *Let $\bar{\epsilon} \in (0, 1)$, $\tilde{F}_{\bar{\epsilon}} \subset B(\theta, M_0)$, $0 < \epsilon_0 \leq \bar{\epsilon}$, $\Lambda \in (0, 2^{-1})$,*

$$0 < \delta_0 \leq 16^{-1}\Lambda^2 \epsilon_0 \eta(6M_0 + 3, \epsilon_0(6M_0 + 3)^{-1})\widehat{\Delta},$$

$$n_0 = \lfloor 8(2M_0 + 1)\widehat{\Delta}^{-1}\epsilon_0^{-1}\Lambda^{-2}\eta^{-1}(6M_0 + 3, \epsilon_0(6M_0 + 3)^{-1})\rfloor + 1.$$

Assume that

$$\{\alpha_i\}_{i=0}^{\infty} \subset (\Lambda, 1 - \Lambda),$$

$$\beta_{n,i} \in [\widehat{\Delta}, 1], \ n = 0, 1, \ldots, \ i = 1, \ldots, m,$$

$$\sum_{i=1}^{m} \beta_{n,i} = 1, \ n = 0, 1, \ldots,$$

$$x_0 \in B(0, M_0) \cap C$$

and that for each integer $n \geq 0$,

$$x_{n+1} \in C \cap B((1 - \alpha_n)x_n + \alpha_n \sum_{i=1}^{n} \beta_{n,i} T_i(x_n), \delta_0).$$

Then there exists a strictly increasing sequence of integers $\{n_k\}_{k=1}^{\infty}$ such that

$$1 \leq n_1 \leq n_0$$

and that for each integer $k \geq 1$, $1 \leq n_{k+1} - n_k \leq n_0$ and $x_{n_k} \in F_{\epsilon_0}$.

Theorem 5.10 *Assume that $\bar{\epsilon} \in (0, 1)$, $\epsilon \in (0, \bar{\epsilon}]$, $\epsilon_1 = \epsilon/2$,*

$$F_{\bar{\epsilon}} \subset B(\theta, M_0), \tag{5.95}$$

$$\Lambda \in (0, 2^{-1}),$$

$$\|T_i(x) - T_i(y)\| \leq \|x - y\|, \ x, y \in C, \ i = 1, \ldots, m, \tag{5.96}$$

$$\epsilon_0 = 8^{-1}\epsilon_1 \widehat{\Delta} \eta(6M_0 + 1, 2^{-1}\epsilon_1(6M_0 + 1)^{-1}), \tag{5.97}$$

$$N_0 = \lfloor 8M_0\epsilon_0^{-1}\Lambda^{-2}\eta^{-1}(6M_0 + 1, \epsilon_0(6M_0 + 1)^{-1})\widehat{\Delta}^{-1} \rfloor + 1, \tag{5.98}$$

$$\delta = (8N_0)^{-1}\epsilon, \tag{5.99}$$

$$\alpha \in (\Lambda, 1 - \Lambda), \tag{5.100}$$

$$\beta_i \in [\widehat{\Delta}, 1], \ i = 1, \ldots, m, \ \sum_{i=1}^{m} \beta_i = 1, \tag{5.101}$$

$$x_0 \in B(0, M_0) \cap C \tag{5.102}$$

and that for each integer $n \geq 0$,

$$x_{n+1} \in B((1-\alpha)x_n + \alpha \sum_{i=1}^{m} \beta_i T_i(x_n), \delta) \cap C. \qquad (5.103)$$

Then for each integer $n \geq N_0$, $x_n \in F_\epsilon$.

Proof Define a sequence $\{y_i\}_{i=0}^{\infty} \subset C$ such that

$$y_0 = x_0 \qquad (5.104)$$

and that for each integer $n \geq 0$,

$$y_{n+1} = (1-\alpha)y_n + \alpha \sum_{i=1}^{m} \beta_i T_i(y_n). \qquad (5.105)$$

Proposition 5.1 and equations (5.102), (5.104) and (5.105) imply that for each integer $n \geq 0$,

$$\|y_n\| \leq 3M_0. \qquad (5.106)$$

Theorem 5.5 and equations (5.96)–(5.98), (5.100)–(5.102), (5.104), and (5.105) imply that for each integer $n \geq N_0$,

$$y_n \in F_{\epsilon/2}. \qquad (5.107)$$

We show that for each integer $n \geq 0$,

$$\|x_n - y_n\| \leq n\delta. \qquad (5.108)$$

Clearly, (5.108) holds for $n = 0$. Assume that $k \geq 0$ is an integer and (5.108) holds for $n = k$. Equations (5.92), (5.101), (5.103), (5.105), and (5.108) imply that

$$\|x_{k+1} - y_{k+1}\| \leq \|x_{k+1} - (1-\alpha)x_k - \alpha \sum_{i=1}^{m} \beta_i T_i(x_k)\|$$

$$+ \|(1-\alpha)x_k + \alpha \sum_{i=1}^{m} \beta_i T_i(x_k) - (1-\alpha)y_k - \alpha \sum_{i=1}^{m} \beta_i T_i(y_k)\|$$

$$\leq \delta + \|x_k - y_k\| \leq (k+1)\delta$$

and (5.108) holds for $n = k + 1$ too. Thus by induction we showed that (5.108) holds for each integer $n \geq 0$.

Assume that $n \in \{N_0, \ldots, 2N_0\}$. By (5.96), (5.99), (5.107), and (5.108),

$$\|x_n - y_n\| \leq 2N_0\delta$$

and for each $s \in \{1, \ldots, m\}$,

$$\|x_n - T_s(x_n)\| \leq \|x_n - y\| + \|y_n - T_s(y_n)\| + \|T_s(y_n) - T_s(x_n)\|$$

$$\leq 2^{-1}\epsilon + 2\|x_n - y_n\| \leq \epsilon/2 + 4N_0\delta \leq \epsilon$$

and $x_n \in F_\epsilon, n = N_0, \ldots, 2N_0$.

Assume that $q \geq 0$ is an integer and

$$x_i \in F_\epsilon, \ i \in \{q, \ldots, q + N_0\}. \tag{5.109}$$

Define a sequence $\{y_i\}_{i=q}^\infty \subset C$ such that

$$y_q = x_q \tag{5.110}$$

and that for each integer $n \geq q$,

$$y_{n+1} = (1 - \alpha)y_n + \alpha \sum_{i=1}^m \beta_i T_i(y_n). \tag{5.111}$$

Equations (5.95), (5.109), and (5.110) imply that

$$\|y_q\| = \|x_q\| \leq M_0.$$

Consider the sequence $\{y_{q+i}\}_{i=0}^\infty$. Theorem 5.5, equations (5.96)–(5.98), (5.100), (5.101), and the equation above imply that for each integer $i \geq N_0$,

$$y_{q+i} \in F_{\epsilon/2}. \tag{5.112}$$

We show that for each integer $n \geq 0$,

$$\|x_{q+n} - y_{q+n}\| \leq n\delta. \tag{5.113}$$

In view of (5.110), equation (5.113) holds for $n = 0$. Assume that $n \geq 0$ is an integer and (5.113) holds. Equations (5.96), (5.100), (5.101), (5.103), (5.111), (5.113) imply that

$$\|x_{q+n+1} - y_{q+n+1}\| \leq \|x_{q+n+1} - (1-\alpha)x_{q+n} - \alpha \sum_{i=1}^{m} \beta_i T_i(x_{q+n})\|$$

$$+ \|(1-\alpha)x_{q+n} + \alpha \sum_{i=1}^{m} \beta_i T_i(x_{q+n}) - (1-\alpha)y_{q+n} - \alpha \sum_{i=1}^{m} \beta_i T_i(y_{q+n})\|$$

$$\leq \delta + (1-\alpha)\|x_{q+n} - y_{q+n}\| + \alpha \sum_{i=1}^{m} \beta_i \|T_i(x_{q+n}) - T_i(y_{q+n})\|$$

$$\leq \delta + \|x_{q+n} - x_{q+n}\| \leq \delta(n+1),$$

and our assumption holds for $n + 1$ too. Thus by induction we showed that (5.113) holds for each integer $n \geq 0$.

Assume that $n \in \{N_0, \ldots, 2N_0\}$, $s \in \{1, \ldots, m\}$. By (5.99), (5.112), (5.113),

$$\|x_{q+n} - T_s(x_{q+n})\|$$

$$\leq \|x_{q+n} - y_{q+n}\| + \|y_{q+n} - T_s(y_{q+n})\| + \|T_s(y_{q+n}) - T_s(x_{q+n})\|$$

$$\leq 2^{-1}\epsilon + 2N_0\delta \leq \epsilon$$

and

$$x_{q+n} \in F_\epsilon, \quad n = N_0, \ldots, 2N_0.$$

Thus if $q \geq 0$ is an integer and (5.109) holds, then

$$x_n \in F_\epsilon, \quad n = q + N_0, \ldots, q + 2N_0.$$

This implies that $x_n \in F_\epsilon$ for each integer $n \geq N_0$. Theorem 5.10 is proved.

Chapter 6
Dynamic String-Averaging Methods

Abstract In this chapter we study the convergence of dynamic string-averaging methods for solving common fixed point problems in a normed space. Our main goal is to obtain an approximate solution of the problem in the presence of computational errors. We show that our dynamic string-averaging algorithm generates a good approximate solution, if the sequence of computational errors is bounded from above by a constant. Moreover, for a known computational error, we find out what an approximate solution can be obtained and how many iterates one needs for this.

6.1 Preliminaries

In this chapter we assume that W-hyperbolic space $X = (X, d, W)$ equipped with the structure (X, η), where $\eta : (0, \infty) \times (0, 2] \to (0, 1]$ is the modulus of uniform convexity, is also a normed space $(X, \| \cdot \|)$ and that

$$d(x, y) = \|x - y\|, \ x, y \in X$$

and that for all $\alpha \in [0, 1]$, $x, y \in X$,

$$(1 - \alpha)x \oplus \alpha y = (1 - \alpha)x + \alpha y.$$

Assume that $C \subset X$ is a non-empty convex set, m is a natural number, and that for $i = 1, \ldots, m$, $T_i : C \to C$. Assume that $M_0 > 0$ and the following assumption holds:

(A) For each $\delta \in (0, 1)$ there exists $p_\delta \in B(0, M_0) \cap C$ such that for each $i \in \{1, \ldots, m\}$ and each $x \in C$,

$$\|T_i(x) - p_\delta\| \leq \|x - p_\delta\| + \delta.$$

For each $\delta \in (0, 1)$ let p_δ be as guaranteed by (A).

For each $i \in \{1, \ldots, m\}$ and each $\gamma \in (0, 1)$ set

$$T_{i,\gamma}(x) = (1 - \gamma)x + \gamma T_i(x), \ x \in C. \tag{6.1}$$

Recall that

$$F = \cap_{i=1}^{m}\mathrm{Fix}(T_i)$$

and set for every positive number ϵ and every integer $i \in \{1, \ldots, m\}$,

$$F_\epsilon(T_i) = \{x \in X : \|x - T_i(x)\| \leq \epsilon\},$$

$$\tilde{F}_\epsilon(T_i) = F_\epsilon(T_i) + B(0, \epsilon),$$

$$F_\epsilon = \cap_{i=1}^{m}F_\epsilon(T_i), \ \tilde{F}_\epsilon = \cap_{i=1}^{m}\tilde{F}_\epsilon(T_i).$$

We apply a dynamic string-averaging method with variable strings and weights in order to obtain a good approximative solution of the common fixed point problem.

Next we describe the dynamic string-averaging method with variable strings and weights.

By an index vector, we a mean a vector $t = (t_1, \ldots, t_p)$ such that $t_i \in \{1, \ldots, m\}$ for all $i = 1, \ldots, p$.

For an index vector $t = (t_1, \ldots, t_q)$ and a vector $\alpha = (\alpha_1, \ldots, \alpha_m) \in (0, 1)^m$, set

$$p(t) = q, \ T[t, \alpha] = T_{t_q,\alpha_{t_q}} \cdots T_{t_1,\alpha_{t_1}}. \tag{6.2}$$

Denote by $\mathcal{M}$ the collection of all pairs (Ω, w), where Ω is a finite set of index vectors and

$$w : \Omega \to (0, \infty) \text{ satisfies } \sum_{t \in \Omega} w(t) = 1. \tag{6.3}$$

Let $(\Omega, w) \in \mathcal{M}$ and $\alpha = (\alpha_1, \ldots, \alpha_m) \in (0, 1)^m$. Define

$$T_{\Omega,w,\alpha}(x) = \sum_{t \in \Omega} w(t)T[t, \alpha](x), \ x \in C. \tag{6.4}$$

The dynamic string-averaging method with variable strings and variable weights can now be described by the following algorithm.

Initialization: Select an arbitrary point $x_0 \in C$.

Iterative step: Given a current iteration vector x_k, pick

$$(\Omega_k, w_k) \in \mathcal{M}, \ \alpha_k = (\alpha_{k,1}, \ldots, \alpha_{k,m}) \in (0, 1)^m$$

and calculate the next iteration vector x_{k+1} by

$$x_{k+1} = T_{\Omega_k, w_k, \alpha_k}(x_k).$$

Fix a number

$$\Delta \in (0, m^{-1}] \tag{6.5}$$

and an integer

$$\bar{m} \geq m. \tag{6.6}$$

Denote by $\mathcal{M}_*$ the set of all $(\Omega, w) \in \mathcal{M}$ such that

$$p(t) \leq \bar{m} \quad \text{for all } t \in \Omega, \tag{6.7}$$

$$w(t) \geq \widehat{\Delta} \text{ for all } t \in \Omega. \tag{6.8}$$

Fix a natural number $\bar{N}$.

In the studies of the common fixed point problem, the goal is to find a point $x \in F$. In order to meet this goal, we apply an algorithm generated by

$$\{(\Omega_n, w_n)\}_{n=0}^{\infty} \subset \mathcal{M}_*, \ \{\alpha_n\}_{n=0}^{\infty} \in (0, 1)^m$$

such that for each natural number j,

$$\{1, \ldots, m\} \subset \cup_{i=j}^{j+\bar{N}-1}(\cup_{t \in \Omega_i}\{t_1, \ldots, t_{p(t)}\}).$$

This algorithm generates, for any starting point $x_0 \in X$, a sequence $\{x_k\}_{k=0}^{\infty} \subset X$, where

$$x_{k+1} = T_{\Omega_k, w_k, \alpha_k}(x_k).$$

We use the following definitions.

Let $\delta \geq 0, x \in C, \alpha = (\alpha_1, \ldots, \alpha_m) \in (0, 1)^m$ and let $t = (t_1, \ldots, t_{p(t)})$ be an index vector. Define

$$A_0(x, t, \alpha, \delta) = \{(y, \lambda) \in C \times R^1 : \ \text{there is a sequence } \{y_i\}_{i=0}^{p(t)} \subset C \text{ such that}$$

$$y_0 = x \text{ and for all } i = 0, \ldots, p(t) - 1,$$

$$\|y_{i+1} - T_{t_{i+1}, \alpha_{i+1}}(y_i)\| \leq \delta,$$

$$y = y_{p(t)},$$

$$\lambda = \max\{\|y_i - T_{t_{i+1}}(y_i)\| : \ i = 0, \ldots, p(t) - 1\}\}. \tag{6.9}$$

Let $\delta \geq 0$, $x \in X$ and let $(\Omega, w) \in \mathcal{M}$ and $\alpha = (\alpha_1, \ldots, \alpha_m) \in (0, 1)^m$. Define

$$A(x, (\Omega, w), \alpha, \delta) = \{(y, \lambda) \in C \times R^1 : \text{ there exist}$$

$$(y_t, \lambda_t) \in A_0(x, t, \alpha, \delta), \ t \in \Omega \text{ such that}$$

$$\left\| y - \sum_{t \in \Omega} w(t) y_t \right\| \leq \delta, \ \lambda = \max\{\lambda_t : t \in \Omega\}\}. \tag{6.10}$$

6.2   A Basic Lemma

In this chapter our study is based on the following lemma.

Lemma 6.1 *Assume that* $\{(\Omega_i, w_i)\}_{i=0}^{\infty} \subset \mathcal{M}_*$, $\{\Delta_i\}_{i=0}^{\infty} \subset [0, \infty)$,

$$\Delta = \sum_{i=0}^{\infty} \Delta_i, \tag{6.11}$$

for each integer $i \geq 0$, $\alpha_i = (\alpha_{i,1}, \ldots, \alpha_{i,m}) \in (0, 1)^m$, *for each integer* $j \geq 0$,

$$\{1, \ldots, m\} \subset \cup_{i=j}^{j+\bar{N}-1}(\cup_{t \in \Omega_i}\{t_1, \ldots, t_{p(t)}\}), \tag{6.12}$$

$\{x_i\}_{i=0}^{\infty} \subset C$, $\{\lambda_i\}_{i=1}^{\infty} \subset [0, \infty)$,

$$\|x_0\| \leq M_0, \tag{6.13}$$

for each integer $n \geq 0$,

$$(x_{n+1}, \lambda_{n+1}) \in A(x_n, (\Omega_n, w_n), \alpha_n, \Delta_n), \tag{6.14}$$

$$(y_{n,t}, \lambda_{n,t}) \in A_0(x, t, \alpha_n, \Delta_n), \ t \in \Omega_n, \tag{6.15}$$

$$\left\| x_{n+1} - \sum_{t \in \Omega_n} w_n(t) y_{n,t} \right\| \leq \Delta_n, \tag{6.16}$$

$$\lambda_{n+1} = \max\{\lambda_{n,t} : t \in \Omega_n\}, \tag{6.17}$$

for each $t = (t_1, \ldots, t_{p(t)}) \in \Omega_n$, $\{y_j^{(n,t)}\}_{j=0}^{p(t)} \subset C$,

$$y_0^{(n,t)} = x_n, \ y_{p(t)}^{(n,t)} = y_{n,t}, \tag{6.18}$$

for each $i \in \{0, \ldots, p(t) - 1\}$,

$$\|y_{i+1}^{(n,t)} - T_{t_{i+1}, \alpha_{t_{i+1}}}(y_i^{(n,t)})\| \leq \Delta_n, \tag{6.19}$$

$$\lambda_{n,t} = \max \|y_i^{(n,t)} - T_{t_{i+1}}(y_i^{(n,t)})\| : i = 0, \ldots, p(t) - 1\}. \tag{6.20}$$

Then the following assertions hold:

1. *Let $\delta \in (0, 1)$. Then for each integer $n \geq 0$, each $t = (t_1, \ldots, t_{p(t)}) \in \Omega_n$, each $j \in \{0, \ldots, p(t) - 1\}$, and each $i \in \{0, \ldots, p(t)\}$,*

$$\|y_{j+1}^{(n,t)} - p_\delta\| \leq \|y_j^{(n,t)} - p_\delta\| + \delta + \Delta_n,$$

$$\|x_{n+1} - p_\delta\|, \ \|y_i^{(n,t)} - p_\delta\|, \ \|y_{n,t} - p_\delta\| \leq \|p_\delta - x_n\| + (\delta + \Delta_n)\bar{m} + \Delta_n, \tag{6.21}$$

$$\|x_n\| \leq 3M_0 + \Delta(\bar{m} + 1), \ \|y_{n,t}\| \leq 3M_0 + \Delta\bar{m}, \tag{6.22}$$

$$\|y_i^{(n,t)}\| \leq 3M_0 + \Delta\bar{m}. \tag{6.23}$$

2. *Let $\gamma \in (0, 1]$, $\Delta < \infty$, $n \geq 0$ be an integer, $t \in \Omega_n$, $i \in \{0, \ldots, p(t) - 1\}$,*

$$\|y_i^{(n,t)} - T_{t_{i+1}}(y_i^{(n,t)})\| > \gamma, \tag{6.24}$$

$$0 < \delta$$

$$\leq 4^{-1}\gamma\alpha_{n,t_{i+1}}(1 \quad \alpha_{n,t_{i+1}})\eta(6M_0 + 2\Delta(\bar{m}+1) + 1, \gamma(6M_0 + 2\Delta(\bar{m}+1) + 1)^{-1}). \tag{6.25}$$

Then

$$\|y_{i+1}^{(n,t)} - p_\delta\| \leq \|y_i^{(n,t)} - p_\delta\| + \Delta_n$$

$$-4^{-1}\gamma\eta(6M_0 + 2\Delta(\bar{m}+1) + 1, \gamma(6M_0 + 2\Delta(\bar{m}+1) + 1)^{-1})\alpha_{n,t_{i+1}}(1 - \alpha_{n,t_{i+1}}).$$

3. *Let $\gamma \in (0, 1]$, $\Delta < \infty$, $n \geq 0$ be an integer, $t \in \Omega_n$,*

$$\lambda_{t,n} > \gamma, \tag{6.26}$$

$$0 < \delta \leq 4^{-1}\gamma \min\{\alpha_{n,j}(1 - \alpha_{n,j}) : j = 1, \ldots, m\}$$

$$\times \eta(6M_0 + 2\Delta(\bar{m} + 1) + 1, \gamma(6M_0 + 2\Delta(\bar{m} + 1) + 1)^{-1}). \tag{6.27}$$

Then

$$\|y_{n,t} - p_\delta\| \le \|x_n - p_\delta\|$$

$$- 4^{-1} \gamma \eta (6M_0 + 2\Delta(\bar{m} + 1) + 1, \gamma (6M_0 + 2\Delta(\bar{m} + 1) + 1)^{-1})$$

$$\times \min\{\alpha_{n,j}(1 - \alpha_{n,j}) : \ j = 1, \dots, m\} + \bar{m}(\Delta_n + \delta).$$

4. *Let* $\gamma \in (0, 1]$, $\Delta < \infty$, $n \ge 0$ *be an integer,* $\lambda_{n+1} > \gamma$, $\delta > 0$ *satisfy* (6.27). *Then*

$$\|x_{n+1} - p_\delta\| \le \|x_n - p_\delta\| + 2\bar{m}(\delta + \Delta_n) + 2\Delta_n$$

$$- 4^{-1} \gamma \widehat{\Delta} \eta (6M_0 + 2\Delta(\bar{m} + 1) + 1, \gamma (6M_0 + 2\Delta(\bar{m} + 1) + 1)^{-1})$$

$$\times \min\{\alpha_{n,j}(1 - \alpha_{n,j}) : \ j = 1, \dots, m\}.$$

5. *Let* $\gamma \in (0, 1]$, $n \ge 0$ *be an integer,* $\lambda_{k+1} \le \gamma$, $k \in \{n, \dots, n + \bar{N} - 1\}$,

$$\tilde{\epsilon} = (\bar{m} + 1)(\bar{N} + 1)(\gamma + \max\{\Delta_i : \ i = n, \dots, n + \bar{N}\}). \tag{6.28}$$

Then for each $k \in \{n, \dots, n + \bar{N}\}$,

$$x_k \in \tilde{F}_{\tilde{\epsilon}},$$

and for each $k_1, k_2 \in \{n, \dots, n + \bar{N}\}$,

$$\|x_{k_1} - x_{k_2}\| \le \bar{m} \gamma \bar{N} + (\bar{m} + 1)\bar{N} \max\{\Delta_k : \ k = n, \dots, n + \bar{N}\}.$$

Proof Let us prove Assertion 1. Let $n \ge 0$ be an integer, $t = (t_1, \dots, t_{p(t)}) \in \Omega_n$, $i \in \{0, \dots, p(t) - 1\}$. By assumption (A), (6.1), (6.18), (6.19),

$$\|p_\delta - T_{t_{i+1}}(y_i^{(n,t)})\| \le \|p_\delta - y_i^{(n,t)}\| + \delta, \tag{6.29}$$

$$\|p_\delta - y_{i+1}^{(n,t)}\| = \|p_\delta - T_{t_{i+1}, \alpha_{t_{i+1}}}(y_i^{(n,t)})\| + \|T_{t_{i+1}, \alpha_{t_{i+1}}}(y_i^{(n,t)}) - y_{i+1}^{(n,t)}\|$$

$$\le \|p_\delta - T_{t_{i+1}, \alpha_{t_{i+1}}}(y_i^{(n,t)})\| + \Delta_n \le \|p_\delta - y_i^{(n,t)}\| + \delta + \Delta_n. \tag{6.30}$$

By (6.16), (6.18), (6.30), for all $i = 0, \dots, p(t)$,

$$\|p_\delta - y_i^{(n,t)}\| \le \|p_\delta - x_n\| + (\delta + \Delta_n)\bar{m},$$

$$\|p_\delta - y_{n,t}\| \le \|p_\delta - x_n\| + (\delta + \Delta_n)\bar{m},$$

$$\| p_\delta - x_{n+1} \| \le \| p_\delta - \sum_{t \in \Omega_n} w_n(t) y_{n,t} \| + \Delta_n$$

$$\le \Delta_n + \sum_{t \in \Omega_n} w_n(t) \| p_\delta - y_{n,t} \|$$

$$\le \| p_\delta - x_n \| + (\delta + \Delta_n) \bar{m} + \Delta_n.$$

Since δ is any element of the interval $(0, 1)$, using (A), we conclude that for each integer $n \ge 0$, each $t = (t_1, \ldots. t_{p(t)}) \in \Omega_n$, and each $i \in \{1, \ldots, p(t)\}$,

$$\| x_n \| \le 3M_0 + \Delta(\bar{m} + 1), \quad \| y_{n,t} \| \le 3M_0 + \Delta(\bar{m} + 1),$$

$$\| y_i^{(n,t)} \| \le 3M_0 + \Delta(\bar{m} + 1).$$

Assertion 1 is proved.

Let us prove Assertion 2. By assumption (A), equations (6.21), (6.24), (6.25) and Lemma 2.4 applied with $S = T_{t_{i+1}}$, $u = y_i^{(n,t)}$, $\alpha = \alpha_{n,t_{i+1}}$, $M = 3M_0 + \Delta(\bar{m} + 1)$, $p = p_\delta$, and

$$v = (1 - \alpha_{n,t_{i+1}}) y_i^{(n,t)} + \alpha_{n,t_{i+1}} T_{t_{i+1}}(y_i^{(n,t)}),$$

we have

$$\| \alpha_{n,t_{i+1}} T_{t_{i+1}}(y_i^{(n,t)}) + (1 - \alpha_{n,t_{i+1}}) y_i^{(n,t)} - p_\delta \| \le \| y_i^{(n,t)} - p_\delta \|$$

$$- 4^{-1} \gamma \eta (6M_0 + 2\Delta(\bar{m} + 1) + 1, \gamma (6M_0 + 2\Delta(\bar{m} + 1) + 1)^{-1}) \alpha_{n,t_{i+1}} (1 - \alpha_{n,t_{i+1}}).$$

Together with (6.19), this implies that

$$\| y_{i+1}^{(n,t)} - p_\delta \| \le \| y_i^{(n,t)} - \alpha_{n,t_{i+1}} T_{t_{i+1}}(y_i^{(n,t)}) - (1 - \alpha_{n,t_{i+1}}) y_i^{(n,t)} \|$$

$$+ \| \alpha_{n,t_{i+1}} T_{t_{i+1}}(y_i^{(n,t)}) + (1 - \alpha_{n,t_{i+1}}) y_i^{(n,t)} - p_\delta \|$$

$$\le \Delta_n + \| y_i^{(n,t)} - p_\delta \|$$

$$- 4^{-1} \gamma \eta (6M_0 + 2\Delta(\bar{m} + 1) + 1, \gamma (6M_0 + 2\Delta(\bar{m} + 1) + 1)^{-1}) \alpha_{n,t_{i+1}} (1 - \alpha_{n,t_{i+1}}).$$
$$(6.31)$$

Assertion 2 is proved.

Let us prove Assertion 3. In view of (6.20) and (6.26), there exists $i \in \{0, \ldots, p(t) - 1\}$ such that (6.24) holds. Assertions 1 and 2, (6.18), (6.27), and (6.31) imply that

$$\|x_n - p_\delta\| - \|y_{n,t} - p_\delta\| = \|y_0^{(n,t)} - p_\delta\| - \|y_{p(t)}^{(n,t)} - p_\delta\|$$

$$= \sum_{j=0}^{p(t)-1} (\|y_j^{(n,t)} - p_\delta\| - \|y_{j+1}^{(n,t)} - p_\delta\|)$$

$$= -\|y_{i+1}^{(n,t)} - p_\delta\| + \|y_i^{(n,t)} - p_\delta\|$$

$$+ \sum\{\|y_j^{(n,t)} - p_\delta\| - \|y_{j+1}^{(n,t)} - p_\delta\| : \ j \in \{0, \ldots, p(t) - 1\} \setminus \{i\}\}$$

$$\geq 4^{-1}\gamma\eta(6M_0 + 2\Delta(\bar{m} + 1) + 1, \gamma(6M_0 + 2\Delta(\bar{m} + 1) + 1)^{-1})$$

$$\times \alpha_{n,t_{i+1}}(1 - \alpha_{n,t_{i+1}}) - \Delta_n - (\delta + \Delta_n)(p(t) - 1)$$

$$\geq 4^{-1}\gamma\eta(6M_0 + 2\Delta(\bar{m} + 1) + 1, \gamma(6M_0 + 2\Delta(\bar{m} + 1) + 1)^{-1})\alpha_{n,t_{i+1}}(1 - \alpha_{n,t_{i+1}})$$

$$- \Delta_n\bar{m} - \delta\bar{m}.$$

Assertion 3 is proved.

Let us prove Assertion 4. In view of (6.1), there exists $\bar{t} \in \Omega_n$ such that

$$\lambda_{n,\bar{t}} > \gamma.$$

Assertion 3 implies that

$$\|y_{n,\bar{t}} - p_\delta\| \leq \|x_n - p_\delta\|$$

$$- 4^{-1}\gamma\eta(6M_0 + 2\Delta(\bar{m} + 1) + 1, \gamma(6M_0 + 2\Delta(\bar{m} + 1) + 1)^{-1})$$

$$\min\{\alpha_{n,j}(1 - \alpha_{n,j}) : \ j = 1, \ldots, m\} + \bar{m}(\Delta_n + \delta). \tag{6.32}$$

Set

$$\kappa = min\{\alpha_{n,j}(1 - \alpha_{n,j}) : \ j = 1, \ldots, m\}. \tag{6.33}$$

Assertion 1, the convexity of the norm, and equations (6.21) (holding for each $t \in \Omega_n$), (6.16), (6.32), and (6.33) imply that

$$\|x_{n+1} - p_\delta\| \le \|x_{n+1} - \sum_{t \in \Omega_n} w_n(t) y_{n,t}\|$$

$$+ \|\sum_{t \in \Omega_n} w_n(t) y_{n,t} - p_\delta\|$$

$$\le \Delta_n + \sum_{t \in \Omega_n} w_n(t) \|y_{n,t} - p_\delta\|$$

$$\le \Delta + \|x_n - p_\delta\| + \sum_{t \in \Omega_n} w_n(t)(\|y_{n,t} - p_\delta\| - \|x_n - p_\delta\|)$$

$$\le \Delta + \|x_n - p_\delta\|$$

$$+ w_n(\bar{t})(-4^{-1}\gamma\eta(6M_0 + 2\Delta(\bar{m}+1) + 1, \gamma(6M_0 + 2\Delta(\bar{m}+1) + 1)^{-1})\kappa + \bar{m}(\Delta_n + \delta))$$

$$+ (\delta + \Delta_n)\bar{m} + \Delta_n \le 2\Delta_n + \|x_n - p_\delta\|$$

$$- 4^{-1}\widehat{\Delta}\gamma\eta(6M_0 + 2\Delta(\bar{m}+1) + 1, \gamma(6M_0 + 2\Delta(\bar{m}+1) + 1)^{-1})\kappa + 2\bar{m}(\Delta_n + \delta).$$

Assertion 4 is proved.

Let us prove Assertion 5. Assume that

$$k \in \{n, \ldots, n + \bar{N} - 1\}.$$

It follows from our assumptions, (6.17), (6.20), that for each $t = (t_1, \ldots, t_{p(t)}) \in \Omega_k$ and each $\{0, \ldots, p(t) - 1\}$,

$$\|y_i^{(k,t)} - T_{t_{i+1}}(y_i^{(k,t)})\| \le \gamma \tag{6.34}$$

and in view of (6.1) and (6.19),

$$\|y_i^{(k,t)} - y_{i+1}^{(k,t)}\|$$

$$\le \|y_{i+1}^{(k,t)} - (1 - \alpha_{k,t_{i+1}})y_i^{(k,t)} - \alpha_{k,t_{i+1}} T_{t_{i+1}}(y_i^{(k,t)})\|$$

$$+ \|(1 - \alpha_{k,t_{i+1}})y_i^{(k,t)} - \alpha_{k,t_{i+1}} T_{t_{i+1}}(y_i^{(k,t)}) - y_i^{(k,t)}\| \le \Delta_k + \gamma. \tag{6.35}$$

Equations (6.18) and (6.35) imply that for each $t = (t_1, \ldots, t_{p(t)}) \in \Omega_k$ and each $i, j \in \{0, \ldots, p(t)\}$,

$$\|y_i^{(k,t)} - y_j^{(k,t)}\| \leq \bar{m}(\Delta_k + \gamma), \quad \|x_k - y_i^{(k,t)}\|$$

$$\leq \bar{m}(\Delta_k + \gamma), \quad \|x_k - y_{k,t}\| \leq \bar{m}(\Delta_k + \gamma). \tag{6.36}$$

By (6.16), (6.36) and the convexity of the norm,

$$\|x_k - x_{k+1}\| \leq \|x_{k+1} - \sum_{t \in \Omega_k} w_k(t) y_{k,t}\|$$

$$+ \|\sum_{t \in \Omega_k} w_k(t) y_{k,t} - x_k\|$$

$$\leq \Delta_k + \sum_{t \in \Omega_k} w_k(t) \|y_{k,t} - x_k\| \leq \Delta_k + \bar{m}(\Delta_k + \gamma). \tag{6.37}$$

In view of (6.37), for every pair of integers $k_1, k_2 \in \{n, \ldots, n + \bar{N}\}$,

$$\|x_{k_1} - x_{k_2}\| \leq \bar{m}\gamma\bar{N} + (\bar{m} + 1)\bar{N} \max\{\Delta_k : k = n, \ldots, n + \bar{N}\}. \tag{6.38}$$

Let $k \in \{n, \ldots, n + \bar{N}\}$, $s \in \{1, \ldots, m\}$. By (6.12), there exist an integer $l \in \{n, \ldots, n + \bar{N} - 1\}$ and an index vector $t = (t_1, \ldots, t_{p(t)}) \in \Omega_l$ such that

$$s \in \{t_1, \ldots, t_{p(t)}\}.$$

Thus there exists $i \in \{1, \ldots, p(t)\}$ such that $s = t_i$. Together with (6.34), this implies that

$$\|y_{i-1}^{(l,t)} - T_s(y_{i-1}^{(l,t)})\| = \|y_{i-1}^{(l,t)} - T_{t_i}(y_{i-1}^{(l,t)})\| \leq \gamma.$$

By (6.36),

$$\|x_l - y_{i-1}^{(l,t)}\| \leq \bar{m}(\Delta_l + \gamma),$$

$$\|x_k - y_{i-1}^{(l,t)}\| \leq \|x_k - x_l\| + \|x_l - y_{i-1}^{(l,t)}\|$$

$$\leq \bar{m}\gamma(\bar{N} + 1) + (\bar{m} + 1)(\bar{N} + 1) \max\{\Delta_k : k = n, \ldots, n + \bar{N}\}$$

and $x_k \in \tilde{F}_{\tilde{\epsilon}}$. Assumption 5 is proved. This completes the proof of Lemma 6.1.

6.3 Inexact Iterates with Summable Errors

The results of this section describe the behavior of inexact iterates of our algorithm with summable errors.

Theorem 6.2 *Assume that* $\{(\Omega_i, w_i)\}_{i=0}^{\infty} \subset \mathcal{M}_*, \{\Delta_i\}_{i=0}^{\infty} \subset [0, \infty),$

$$\Delta = \sum_{i=0}^{\infty} \Delta_i < \infty, \tag{6.39}$$

for each integer $i \geq 0$, $\alpha_i = (\alpha_{i,1}, \ldots, \alpha_{i,m}) \in (0, 1)^m$, *for each integer* $j \geq 0$,

$$\{1, \ldots, m\} \subset \bigcup_{i=j}^{j+\bar{N}-1}(\cup_{t \in \Omega_i}\{t_1, \ldots, t_{p(t)}\}), \tag{6.40}$$

$\{x_i\}_{i=0}^{\infty} \subset C, \{\lambda_i\}_{i=1}^{\infty} \subset [0, \infty),$

$$\|x_0\| \leq M_0, \tag{6.41}$$

for each integer $n \geq 0$,

$$(x_{n+1}, \lambda_{n+1}) \in A(x_n, (\Omega_n, w_n), \alpha_n, \Delta_n), \tag{6.42}$$

$\epsilon \in (0, 1),$

$$\epsilon_0 = \epsilon(\bar{N} + 1)^{-1}(2\bar{m} + 2)^{-1}, \tag{6.43}$$

n_0, Q *are natural numbers such that*

$$\Delta_i \leq \epsilon_0 \text{ for each } i \geq n_0\bar{N}, \tag{6.44}$$

$$\sum_{n=n_0}^{n_0+Q} \min\{\alpha_{k,j}(1 - \alpha_{k,j}) : j = 1, \ldots, m, \ k = n\bar{N}, \ldots, (n+1)\bar{N} - 1)\}$$

$$> 4(4M_0 + (3\bar{m} + 1)\Delta)\epsilon_0^{-1}\widehat{\Delta}^{-1}$$

$$\times \eta^{-1}(6M_0 + 2\Delta(\bar{m} + 1) + 1, \epsilon_0(6M_0 + 1 + 2\Delta(\bar{m} + 1))^{-1}). \tag{6.45}$$

Then there exists an integer $n \in \{n_0, \ldots, n_0 + Q\}$ *such that*

$$\lambda_{k+1} \leq \epsilon_0, \ k = n\bar{N}, \ldots, (n+1)\bar{N} - 1. \tag{6.46}$$

Moreover if $n \in \{n_0, \ldots, n_0 + Q\}$ *and* (6.46) *holds, then*

$$x_k \in \tilde{F}_\epsilon, \ k = n\bar{N}, \ldots, (n+1)\bar{N}.$$

Proof For each integer $n \geq 0$, each $t = (t_1, \ldots, t_{p(t)}) \in \Omega_n$, and each $i \in \{0, \ldots, p(t)\}$, define $y_{n,t}$, $y_i^{(n,t)}$, $\lambda_{n,t}$ as in Lemma 6.1 such that (6.14)–(6.20) hold. Choose a positive number

$$\delta \leq 8^{-1}\epsilon_0 \alpha_{n,i}(1 - \alpha_{n,i})\eta(6M_0 + 2\Delta(\bar{m} + 1) + 1, \epsilon_0(6M_0 + 2\Delta(\bar{m} + 1) + 1)^{-1})$$
$$(6.47)$$

for each $i \in \{1, \ldots, m\}$ and each $n \in \{0, \ldots, (n_0 + Q + 2)\bar{N}\}$. We show that there exists $n \in \{n_0, \ldots, n_0 + Q\}$ such that

$$\lambda_{k+1} \leq \epsilon_0, \ k = n\bar{N}, \ldots, (n+1)\bar{N} - 1.$$

Assume the contrary. Then for each $n \in \{n_0, \ldots, n_0 + Q\}$,

$$\max\{\lambda_{k+1} : \ k = n\bar{N}, \ldots, (n+1)\bar{N} - 1\} > \epsilon_0. \tag{6.48}$$

Let $n \in \{n_0, \ldots, n_0 + Q\}$. In view of (6.48), there exists

$$k \in \{n\bar{N}, \ldots, (n+1)\bar{N} - 1\}$$

such that

$$\lambda_{k+1} > \epsilon_0. \tag{6.49}$$

Assertion 4 of Lemma 6.1 with $\gamma = \epsilon_0$ and (6.39) imply that

$$\|x_{k+1} - p_\delta\| \leq \|x_k - p_\delta\| + 2\bar{m}(\delta + \Delta_k) + 2\Delta_k$$

$$- 4^{-1}\epsilon_0\widehat{\Delta}\eta(6M_0 + 2\Delta(\bar{m} + 1) + 1, \epsilon_0(6M_0 + 2\Delta(\bar{m} + 1) + 1)^{-1})$$

$$\times \min\{\alpha_{k,j}(1 - \alpha_{k,j}) : \ j = 1, \ldots, m\}. \tag{6.50}$$

Assertion 1 of Lemma 6.1 implies that for each $i \in \{n\bar{N}, \ldots, (n+1)\bar{N} - 1\}$,

$$\|x_{i+1} - p_\delta\| \leq \|p_\delta - x_i\| + (\delta + \Delta_i)\bar{m} + \Delta_i. \tag{6.51}$$

Equations (6.50) and (6.51) imply that

$$\|x_{n\bar{N}} - p_\delta\| - \|x_{(n+1)\bar{N}} - p_\delta\|$$

$$= \sum_{i=n\bar{N}}^{(n+1)\bar{N}-1} (\|x_i - p_\delta\| - \|x_{i+1} - p_\delta\|)$$

$$\geq 4^{-1}\epsilon_0\widehat{\Delta}\eta(6M_0+2\Delta(\bar{m}+1)+1, \epsilon_0(6M_0+2\Delta(\bar{m}+1)+1)^{-1})\min\{\alpha_{i,j}(1-\alpha_{i,j}) :$$

$$j = 1,\ldots,m, \ i = n\bar{N},\ldots,(n+1)\bar{N}-1\} - 2\delta\bar{m}\bar{N} - 2\bar{m}\sum_{i=n\bar{N}}^{(n+1)\bar{N}-1}\Delta_i. \quad (6.52)$$

Assertion 1 of Lemma 6.1 and assumption (A) imply that

$$4M_0 + \Delta(\bar{m} + 1) \geq \|x_{n_0\bar{N}} - p_\delta\| \geq \|x_{n_0\bar{N}} - p_\delta\| - \|x_{(Q+1+n_0)\bar{N}} - p_\delta\|$$

$$= \sum_{i=n_0}^{Q+n_0} (\|x_{n\bar{N}} - p_\delta\| - \|x_{(n+1)\bar{N}} - p_\delta\|)$$

$$\geq 4^{-1}\epsilon_0\widehat{\Delta}\eta(6M_0 + 2\Delta(\bar{m} + 1) + 1, \epsilon_0(6M_0 + 2\Delta(\bar{m} + 1) + 1)^{-1})$$

$$\times \sum_{n=n_0}^{Q+n_0} \min\{\alpha_{i,j}(1 - \alpha_{i,j}) :$$

$$j = 1,\ldots,m, \ i = n\bar{N},\ldots,(n+1)\bar{N}-1\} - 2\delta\bar{m}\bar{N}(Q+1) - 2\bar{m}\sum_{i=n_0\bar{N}}^{(Q+1+n_0)\bar{N}-1}\Delta_i.$$

Since δ is any positive number satisfying (6.47), we conclude that

$$\sum_{n=n_0}^{Q+n_0} \min\{\alpha_{i,j}(1 - \alpha_{i,j}) :$$

$$j = 1,\ldots,m, \ i = n\bar{N},\ldots,(n + 1)\bar{N} - 1\}$$

$$\leq 4(4M_0 + \Delta(3\bar{m} + 1))\epsilon_0^{-1}\widehat{\Delta}^{-1}$$

$$\times \eta^{-1}(6M_0 + 2\Delta(\bar{m} + 1) + 1, \epsilon_0(6M_0 + 2\Delta(\bar{m} + 1) + 1)^{-1}).$$

This contradicts (6.45). The contradiction we have reached proves that there exists $n \in \{n_0, \ldots, n_0 + Q\}$ such that

$$\lambda_{k+1} \leq \epsilon_0, \ \ k = n\bar{N}, \ldots, (n+1)\bar{N} - 1. \tag{6.53}$$

Assume that $n \in \{n_0, \ldots, n_0 + Q\}$ and (6.53) holds. Assertion 5 of Lemma 6.1 applied with $\gamma = \epsilon_0$ and (6.43), (6.44) imply that for each $k \in \{n\bar{N}, \ldots, (n+1)\bar{N}\}$, $x_k \in \tilde{F}_{\tilde{\epsilon}}$, where

$$\tilde{\epsilon} = (\bar{m} + 1)(\bar{N} + 1)(\epsilon_0 + \max\{\Delta_i : \ i = n\bar{N}, \ldots, (n+1)\bar{N}\})$$

$$\leq 2(\bar{m} + 1)(\bar{N} + 1)\epsilon_0 = \epsilon.$$

Theorem 6.2 is proved.

Theorem 6.2 implies the following result.

Theorem 6.3 *Assume that the assumptions of Theorem 6.2 hold and*

$$\sum_{n=0}^{\infty} \min\{\alpha_{k,j}(1 - \alpha_{k,j}) : \ j = 1, \ldots, m, \ k = n\bar{N}, \ldots, (n+1)\bar{N} - 1)\} = \infty.$$

Then there exists a strictly increasing sequence of natural numbers $\{n_k\}_{k=1}^{\infty}$ such that for each integer $k \geq 1$ and each $i \in \{n_k, \bar{N}, \ldots, (n_k + 1)\bar{N}\}$, $x_i \in \tilde{F}_{1/k}$.

Theorem 6.4 *Assume that $\Lambda \in (0, 2^{-1})$,*

$$\{(\Omega_i, w_i)\}_{i=0}^{\infty} \subset \mathcal{M}_*, \tag{6.54}$$

$\{\Delta_i\}_{i=0}^{\infty} \subset [0, \infty)$,

$$\Delta = \sum_{i=0}^{\infty} \Delta_i < \infty, \tag{6.55}$$

for each integer $i \geq 0$,

$$\alpha_i = (\alpha_{i,1}, \ldots, \alpha_{i,m}) \in (\Lambda, 1 - \Lambda)^m, \tag{6.56}$$

for each integer $j \geq 0$,

$$\{1, \ldots, m\} \subset \cup_{i=j}^{j+\bar{N}-1}(\cup_{t \in \Omega_i}\{t_1, \ldots, t_{p(t)}\}), \tag{6.57}$$

$\{x_i\}_{i=0}^{\infty} \subset C$, $\{\lambda_i\}_{i=1}^{\infty} \subset [0, \infty)$,

$$\|x_0\| \leq M_0, \tag{6.58}$$

for each integer $n \geq 0$,

$$(x_{n+1}, \lambda_{n+1}) \in A(x_n, (\Omega_n, w_n), \alpha_n, \Delta_n), \qquad (6.59)$$

$\epsilon \in (0, 1)$,

$$\epsilon_0 = \epsilon(\bar{N} + 1)^{-1}(2\bar{m} + 2)^{-1}, \qquad (6.60)$$

and n_0 is a natural number such that

$$\Delta_i \leq \epsilon_0/2 \text{ for each } i \geq n_0. \qquad (6.61)$$

Then

$$Card(\{n \in \{0, 1, \ldots\} : \max\{\lambda_{k+1} : k \in \{n, \ldots, n + \bar{N} - 1\} > \epsilon_0\})$$

$$\leq 4\bar{N}(2M_0 + (2\bar{m} + 2)\Delta)\epsilon_0^{-1}\widehat{\Delta}^{-1}\Lambda^{-2}$$

$$\times \eta^{-1}(6M_0 + 2\Delta(\bar{m} + 1) + 1, \epsilon_0(6M_0 + 1 + 2\Delta(\bar{m} + 1))^{-1}).$$

Moreover if an integer $n \geq n_0$ and

$$\lambda_{k+1} \leq \epsilon_0, \ k \in \{n, \ldots, n + \bar{N} - 1\},$$

then

$$x_k \in \tilde{F}_\epsilon, \ k = n, \ldots, n + \bar{N}.$$

Proof For each integer $n \geq 0$, each $t = (t_1, \ldots, t_{p(t)}) \in \Omega_n$, and each $i \in \{0, \ldots, p(t)\}$, define $y_{n,t}, y_i^{(n,t)}, \lambda_{n,t}$ as in Lemma 6.1 such that (6.14)–(6.20) hold. Let $Q > n_0$ be a natural number. Choose a positive number

$$\delta < 8^{-1}\epsilon_0\Lambda^2\eta(6M_0 + 2\Delta(\bar{m} + 1) + 1, \epsilon_0(6M_0 + 2\Delta(\bar{m} + 1) + 1)^{-1}). \qquad (6.62)$$

Set

$$E = \{n \in \{0, 1, \ldots\} : \lambda_{n+1} > \epsilon_0\}, \qquad (6.63)$$

$$E_Q = E \cap \{0, 1, \ldots, Q\}, \qquad (6.64)$$

$$\tilde{\Delta}_1 = \Delta(\bar{m} + 1). \qquad (6.65)$$

Assume that $n \geq 0$ is an integer and

$$\lambda_{n+1} > \epsilon_0. \tag{6.66}$$

Assertion 4 of Lemma 6.1 with $\gamma = \epsilon_0$ and (6.14), (6.20), (6.56), (6.62), and (6.65) imply that

$$\|x_{n+1} - p_\delta\| \leq \|x_n - p_\delta\| + 2\bar{m}(\delta + \Delta_n) + 2\Delta_n$$

$$- 4^{-1}\epsilon_0 \widehat{\Delta} \Lambda^2 \eta(6M_0 + 2\tilde{\Delta}_1 + 1, \epsilon_0(6M_0 + 2\tilde{\Delta}_1 + 1)^{-1}). \tag{6.67}$$

Assertion 1 of Lemma 6.1, assumption (A), (6.58), and (6.67) imply that

$$2M_0 \geq \|p_\delta - x_0\| \geq \|p_\delta - x_0\| - \|p_\delta - x_{Q+1}\|$$

$$= \sum_{n=0}^{Q} (\|x_n - p_\delta\| - \|x_{n+1} - p_\delta\|)$$

$$= \sum \{\|x_n - p_\delta\| - \|x_{n+1} - p_\delta\| : n \in E_Q\}$$

$$+ \sum \{\|x_n - p_\delta\| - \|x_{n+1} - p_\delta\| : n \in \{0, \ldots, Q\} \setminus E_Q\}$$

$$\geq 4^{-1} \widehat{\Delta} \epsilon_0 \Lambda^2 \eta(6M_0 + 2\tilde{\Delta}_1 + 1, \epsilon_0(6M_0 + 2\tilde{\Delta}_1 + 1)^{-1}) \mathrm{Card}(E_Q)$$

$$- 2\bar{m}\delta \mathrm{Card}(E_Q) - (2\bar{m} + 1) \sum \{\Delta_n : n \in E_Q\}$$

$$- \sum \{(\delta + \Delta_n)\bar{m} + \Delta_n : n \in \{0, \ldots, Q\} \setminus E_Q\}.$$

Since δ is any positive sufficient small number, the relation above and (6.62) imply that

$$\mathrm{Card}(E_Q) \leq 4(2M_0 + (2\bar{m} + 1)\Delta)\widehat{\Delta}^{-1}\epsilon_0^{-1}\Lambda^{-2}$$

$$\times \eta^{-1}(6M_0 + 1 + 2\tilde{\Delta}_1, \epsilon_0(6M_0 + 2\tilde{\Delta}_1 + 1)^{-1}).$$

Since Q is any natural number, we conclude that

$$\mathrm{Card}(E) \leq 4(2M_0 + (2\bar{m} + 1)\Delta)\widehat{\Delta}^{-1}\epsilon_0^{-1}\Lambda^{-2}$$

$$\times \eta^{-1}(6M_0 + 1 + 2\tilde{\Delta}_1, \epsilon_0(6M_0 + 2\tilde{\Delta}_1 + 1)^{-1}). \tag{6.68}$$

Set

$$E_0 = \{n \in \{0, 1, \ldots\} : [n, n + \bar{N} - 1] \cap E \neq \emptyset\}. \tag{6.69}$$

By (6.68) and (6.69),

$$\text{Card}(E_0) \leq \bar{N}\text{Card}(E)$$

$$\leq 4\bar{N}(2M_0 + (2\bar{m} + 1)\Delta)\widehat{\Delta}^{-1}\epsilon_0^{-1}\Lambda^{-2}$$

$$\times \eta^{-1}(6M_0 + 1 + 2\tilde{\Delta}_1, \epsilon_0(6M_0 + 2\tilde{\Delta}_1 + 1)^{-1}).$$

Assume that $n \geq n_0$ is an integer and $n \notin E_0$. In view of (6.63) and (6.69),

$$[n, n + \bar{N} - 1] \cap E = \emptyset,$$

and for each $k \in \{n, \ldots, n + \bar{N} - 1\}$,

$$\lambda_{k+1} \leq \epsilon_0.$$

Assertion 5 of Lemma 6.1 applied with $\gamma = \epsilon_0$ and (6.60), (6.61) imply that for each $k \in \{n, \ldots, n + \bar{N}\}$, $x_k \in \bar{F}_{\tilde{\epsilon}}$, where

$$\tilde{\epsilon} = (\bar{m} + 1)(\bar{N} + 1)(\epsilon_0 + \max\{\Delta_i : i = n, \ldots, n + \bar{N}\})$$

$$\leq 2(\bar{m} + 1)(\bar{N} + 1)\epsilon_0 = \epsilon.$$

Theorem 6.4 is proved.

Theorem 6.4 implies the following result.

Theorem 6.5 *Assume that $\Lambda \in (0, 2^{-1})$, $\{(\Omega_i, w_i)\}_{i=0}^{\infty} \subset \mathcal{M}_*$, for each integer $i \geq 0$,*

$$\alpha_i = (\alpha_{i,1}, \ldots, \alpha_{i,m}) \in (\Lambda, 1 - \Lambda)^m,$$

for each integer $j \geq 0$,

$$\{1, \ldots, m\} \subset \cup_{i=j}^{j+\bar{N}-1}(\cup_{t \in \Omega_i}\{t_1, \ldots, t_{p(t)}\}),$$

$\{x_i\}_{i=0}^{\infty} \subset C, \{\lambda_i\}_{i=1}^{\infty} \subset [0, \infty),$

$$\|x_0\| \leq M_0,$$

for each integer $n \geq 0$,

$$(x_{n+1}, \lambda_{n+1}) \in A(x_n, (\Omega_n, w_n), \alpha_n, 0),$$

$\epsilon \in (0, 1)$ *and*

$$\epsilon_0 = \epsilon(\bar{N} + 1)^{-1}(2\bar{m} + 2)^{-1}.$$

Then

$$Card(\{n \in \{0, 1, \ldots\} : \ \max\{\lambda_{k+1} : \ k \in \{n, \ldots, n + \bar{N} - 1\} > \epsilon_0\})$$

$$\leq 8M_0\epsilon_0^{-1}\widehat{\Delta}^{-1}\Lambda^{-2}\eta^{-1}(6M_0 + 1, \epsilon_0(6M_0 + 1)^{-1}).$$

Moreover if an integer $n \geq n_0$ and

$$\lambda_{k+1} \leq \epsilon_0, \ k \in \{n, \ldots, n + \bar{N} - 1\},$$

then

$$x_k \in \tilde{F}_\epsilon, \ k = n, \ldots, n + \bar{N}.$$

6.4   Exact Iterates

The following theorem describes the behavior of exact iterates of our algorithm.

Theorem 6.6 *Assume that $\Lambda \in (0, 2^{-1})$,*

$$\|T_i(x) - T_i(y)\| \leq \|x - y\|, \ x, y \in C, \ i = 1, \ldots, m, \tag{6.70}$$

$$\{(\Omega_i, w_i)\}_{i=0}^{\infty} \subset \mathcal{M}_*,$$

for each integer $i \geq 0$,

$$\alpha_i = (\alpha_{i,1}, \ldots, \alpha_{i,m}) \in (\Lambda, 1 - \Lambda)^m, \tag{6.71}$$

for each integer $n \geq 0$,

$$(\Omega_{n+\bar{N}}, w_{n+\bar{N}}) = (\Omega_n, w_n), \ \alpha_{n+\bar{N}} = \alpha_n, \tag{6.72}$$

for each integer $j \geq 0$,

$$\{1, \ldots, m\} \subset \cup_{i=j}^{j+\bar{N}-1}(\cup_{t \in \Omega_i}\{t_1, \ldots, t_{p(t)}\}), \tag{6.73}$$

$\{x_i\}_{i=0}^{\infty} \subset C$, $\{\lambda_i\}_{i=1}^{\infty} \subset [0, \infty)$,

$$\|x_0\| \leq M_0, $$

for each integer $n \geq 0$,

$$(x_{n+1}, \lambda_{n+1}) \in A(x_n, (\Omega_n, w_n), \alpha_n, 0), \tag{6.74}$$

$$\epsilon \in (0, 1), \ \epsilon_1 = \epsilon(\bar{N}+1)^{-1}(3\bar{m}+3)^{-1}, \tag{6.75}$$

$$\epsilon_0 = 16^{-1}\epsilon_1\Lambda^2\widehat{\Delta}(\bar{m}\bar{N})^{-1}\eta(6M_0+1, \epsilon_1(6M_0+1)^{-1}), \tag{6.76}$$

$$N_0 = \lfloor 16M_0\bar{N}\epsilon_0^{-1}\Lambda^{-2}\widehat{\Delta}^{-1}\eta^{-1}(6M_0+1, \epsilon_0(6M_0+1)^{-1})\rfloor. \tag{6.77}$$

Then for each integer $k \geq (N_0+1)\bar{N}$, $x_k \in F_\epsilon$.

Proof Theorem 6.5 implies that there exists a set $E \subset \{0, 1, \ldots\}$ such that

$$\text{Card}(E_0) \leq N_0, \tag{6.78}$$

for each $n \in \{0, 1, \ldots\} \setminus E$,

$$\lambda_{k+1} \leq \epsilon_0, \ k = n, \ldots, n + \bar{N} - 1, \tag{6.79}$$

for each $l \in \{n, \ldots, n + \bar{N}\}$,

$$x_l \in \tilde{F}_{\epsilon_1} \subset F_{3\epsilon_1}, \tag{6.80}$$

$$\|x_{k_1} - x_{k_2}\| \leq \bar{N}\bar{m}\epsilon_0, \ k_1, k_2 \in \{n, \ldots, n + \bar{N}\}. \tag{6.81}$$

By (6.78) and (6.81), there exists an integer

$$n_0 \in [1, N_0 + 1] \tag{6.82}$$

such that

$$n_0\bar{N} \notin E, \tag{6.83}$$

$$\|x_{n_0\bar{N}} - x_{(n_0+1)\bar{N}}\| \leq \bar{N}\bar{m}\epsilon_0. \tag{6.84}$$

Let $I : X \to X$ be the identity operator. For each index vector $t = (t_1, \ldots, t_{p(t)})$ and each $\alpha = (\alpha_1, \ldots, \alpha_m) \in (0, 1)^m$, define

$$Q(t, \alpha) = (\alpha_{t_{p(t)}} T_{t_{p(t)}} + (1 - \alpha_{t_{p(t)}})I) \circ \cdots \circ (\alpha_{t_1} T_{t_1} + (1 - \alpha_{t_1})I). \qquad (6.85)$$

For each integer $n \geq 0$ set

$$Q_n = \sum_{t \in \Omega_n} w_n(t) Q(t, \alpha_n). \qquad (6.86)$$

In view of (6.72) and (6.86),

$$Q_{n+\bar{N}} = Q_n, \quad = 0, 1, \ldots. \qquad (6.87)$$

It follows from (6.70), (6.85), and (6.87) that for each $\xi, \eta \in C$,

$$\|Q_n(\xi) - Q_n(\eta)\| \leq \|\xi - \eta\|. \qquad (6.88)$$

By (6.74), (6.85), and (6.86), for each integer $n \geq 0$,

$$x_{n+1} = Q_n(x_n). \qquad (6.89)$$

In view of (6.83),

$$x_{(n_0+1)\bar{N}} = Q_{(n_0+1)\bar{N}-1} \cdots Q_{n_0\bar{N}}(x_{n_0\bar{N}}) = \prod_{i=n_0\bar{N}}^{(n_0+1)\bar{N}-1} Q_i(x_{n_0\bar{N}}), \qquad (6.90)$$

and for each integer $k \geq n_0$,

$$x_{(k+1)\bar{N}} = \prod_{i=n_0\bar{N}}^{(n_0+1)\bar{N}-1} Q_i(x_{k\bar{N}}). \qquad (6.91)$$

By (6.82), (6.84), (6.87), and (6.91),

$$\|x_{(n+1)\bar{N}} - x_{n\bar{N}}\| \leq \|x_{(n_0+1)\bar{N}} - x_{n_0\bar{N}}\| \leq \bar{N}\bar{m}\epsilon_0. \qquad (6.92)$$

Choose a positive number

$$\delta \leq 64^{-1}\epsilon_1(\bar{m}\bar{N})^{-1}\widehat{\Delta}\Lambda^2\eta(6M_0 + 1, \epsilon_1(6M_0 + 1)^{-1}). \qquad (6.93)$$

For each integer $n \geq 0$, each $t = (t_1, \ldots, t_{p(t)}) \in \Omega_n$, and each $i \in \{0, \ldots, p(t)\}$, define $y_{n,t}, y_i^{(n,t)}, \lambda_{n,t}$ as in Lemma 6.1 such that (6.14)–(6.20) hold with $\Delta_n = 0$.

Assertion 1 of Lemma 6.1 implies that for each integer $n \geq 0$, each $t = (t_1, \ldots .t_{p(t)}) \in \Omega_n$, each $j \in \{0, \ldots, p(t) - 1\}$, and each $i \in \{0, \ldots, p(t)\}$,

$$\|y_{j+1}^{(n,t)} - p_\delta\| \leq \|y_j^{(n,t)} - p_\delta\| + \delta, \quad \|y_i^{(n,t)} - p_\delta\| \leq \|x_n - p_\delta\| + \bar{m}\delta, \qquad (6.94)$$

$$\|x_{n+1} - p_\delta\| \leq \|x_n - p_\delta\| + \bar{m}\delta, \qquad (6.95)$$

$$\|x_n - p_\delta\| \leq \|p_\delta - x_0\| + n\delta\bar{m}, \qquad (6.96)$$

$$\|y_i^{(n,t)} - p_\delta\| \leq \|x_0 - p_\delta\| + \bar{m}\delta n, \qquad (6.97)$$

$$\|y_{n,t}\| \leq 3M_0 + \bar{m}\delta n + \bar{m}\delta,$$

$$\|x_n\| \leq 3M_0, \quad \|y_{n,t}\| \leq 3M_0, \quad t \in \Omega_n,$$

$$\|y_i^{(n,t)}\| \leq 3M_0, \quad t \in \Omega_n, \quad i \in \{0, \ldots, p(t)\}. \qquad (6.98)$$

Assume that $n \geq 0$ is an integer, $t = (t_1, \ldots .t_{p(t)}) \in \Omega_n, i \in \{0, \ldots, p(t) - 1\}$, and

$$\|y_i^{(n,t)} - T_{t_{i+1}}(y_i^{(n,t)})\| > \epsilon_1.$$

Assertion 2 of Lemma 6.1 with $\gamma = \epsilon_1$ and (6.71) and (6.93) imply that

$$\|y_{i+1}^{(n,t)} - p_\delta\| \leq \|y_i^{(n,t)} - p_\delta\|$$

$$- 4^{-1}\epsilon_1\eta(6M_0 + 1, \epsilon_1(6M_0 + 1)^{-1})\Lambda^2. \qquad (6.99)$$

Thus the following property holds:

(a) if $n \geq 0$ is an integer, $t = (t_1, \ldots .t_{p(t)}) \in \Omega_n, i \in \{0, \ldots, p(t) - 1\}$ and

$$\|y_i^{(n,t)} - T_{t_{i+1}}(y_i^{(n,t)})\| > \epsilon_1,$$

then (6.99) holds.

Let $n \geq n_0$ be an integer. By (6.92),

$$\epsilon_0 \bar{N}\bar{m} \geq \|x_{(n+1)\bar{N}} - x_{\bar{N}n}\| \geq \|p_\delta - x_{\bar{N}n}\| - \|p_\delta - x_{\bar{N}(n+1)}\|$$

$$= \sum_{k=\bar{N}n}^{(n+1)\bar{N}-1} (\|p_\delta - x_k\| - \|p_\delta - x_{k+1}\|).$$

Together with (6.95), this implies that for each $k \in \{n\bar{N}, \ldots, (n+1)\bar{N} - 1\}$,

$$\|p_\delta - x_k\| - \|p_\delta - x_{k+1}\| \leq \epsilon_0 \bar{m} \bar{N} + \bar{m} \delta \bar{N}. \qquad (6.100)$$

Let

$$k \in \{\bar{N}n, \ldots, (n+1)\bar{N} - 1\}. \qquad (6.101)$$

It follows from (6.16) with $\Delta_n = 0$, (6.100), (6.101) and the convexity of the norm that

$$\bar{m}\bar{N}(\epsilon_0 + \delta) \geq \|p_\delta - x_k\| - \|p_\delta - x_{k+1}\|$$

$$= \|p_\delta - x_k\| - \|p_\delta - \sum_{t \in \Omega_k} w_k(t) y_{k,t}\|$$

$$\geq \|p_\delta - x_k\| - \sum_{t \in \Omega_k} w_k(t)\|p_\delta - y_{k,t}\| \geq \sum_{t \in \Omega_k} w_k(t)(\|p_\delta - x_k\| - \|p_\delta - y_{k,t}\|). \qquad (6.102)$$

In view of (6.94) and (6.102), for each $t \in \Omega_k$,

$$w_k(t)(\|p_\delta - x_k\| - \|p_\delta - y_{k,t}\|) \leq \bar{m}\delta + \bar{m}\bar{N}(\epsilon_0 + \delta),$$

$$\|p_\delta - x_k\| - \|p_\delta - y_{k,t}\| \leq \widehat{\Delta}^{-1}(\bar{m}\bar{N}(\epsilon_0 + \delta) + \bar{m}\delta). \qquad (6.103)$$

Let $t = (t_1, \ldots, t_{p(t)}) \in \Omega_k$. In view of (6.103),

$$\widehat{\Delta}^{-1}(\bar{m}\bar{N}(\epsilon_0 + \delta) + \bar{m}\delta) \geq \|p_\delta - y_0^{(k,t)}\| - \|p_\delta - y_{p(t)}^{(k,t)}\|$$

$$= \sum_{i=0}^{p(t)-1} (\|p_\delta - y_i^{(k,t)}\| - \|p_\delta - y_{i+1}^{(k,t)}\|). \qquad (6.104)$$

By (6.76), (6.93), (6.94), and (6.104), for each integer $i \in \{0, \ldots, p(t) - 1\}$,

$$\|p_\delta - y_i^{(k,t)}\| - \|p_\delta - y_{i+1}^{(k,t)}\| \leq +\bar{m}\delta + \widehat{\Delta}^{-1}(\bar{m}\bar{N}(\epsilon_0 + \delta) + \bar{m}\delta)$$

$$\leq 8^{-1}\epsilon_1\Lambda^2\eta(6M_0 + 1, \epsilon_1(6M_0 + 1)^{-1}).$$

Together with property (a) and (6.99), this implies that for each integer $i \in \{0, \ldots, p(t) - 1\}$,

$$\|y_i^{(k,t)} - T_{t_{i+1}}(y_i^{(k,t)})\| \leq \epsilon_1, \qquad (6.105)$$

which in turn implies that

$$\|y_i^{(k,t)} - y_{i+1}^{(k,t)}\| \le \epsilon_1, \; i = 0, \ldots, p(t) - 1.$$

Thus for each $t = (t_1, \ldots, t_{p(t)}) \in \Omega_k$ and each $i \in \{0, \ldots, p(t) - 1\}$,

$$\|y_i^{(k,t)} - y_{i+1}^{(k,t)}\| \le \epsilon_1,$$

which in turn implies that

$$\|x_k - y_i^{(k,t)}\| \le \bar{m}\epsilon_1, \; i = 0, \ldots, p(t) \tag{6.106}$$

and

$$\|x_k - y_{k,t}\| \le \bar{m}\epsilon_1. \tag{6.107}$$

It follows from (6.74), (6.107) and the convexity of the norm that

$$\|x_k - x_{k+1}\| = \|x_k - \sum_{t \in \Omega_k} w_k(t) y_{k,t}\| \le \bar{m}\epsilon_1. \tag{6.108}$$

By (6.108), for each $k_1, k_2 \in \{n\bar{N}, \ldots, (n+1)\bar{N}\}$,

$$\|x_{k_1} - x_{k_2}\| \le \bar{N}\bar{m}\epsilon_1. \tag{6.109}$$

Let $k \in \{n\bar{N}, \ldots, (n+1)\bar{N}\}$ and $s \in \{1, \ldots, m\}$. In view of (6.73), there exist $k_1 \in \{n\bar{N}, \ldots, (n+1)\bar{N} - 1\}$, $t \in \Omega_{k_1}$ and $i \in \{1, \ldots, p(t)\}$ such that

$$s = t_i. \tag{6.110}$$

By (6.105) and (6.110),

$$\|y_{i-1}^{(k_1,t)} - T_s(y_{i-1}^{(k_1,t)})\| \le \epsilon_1.$$

It follows from (6.75) and (6.109) that

$$\|x_k - y_{i-1}^{(k,t)}\| \le \|x_k - x_{k_1}\| + \|x_{k_1} - y_{i-1}^{(k,t)}\| \le \bar{N}\bar{m}\epsilon_1 + \bar{m}\epsilon_1 \le \epsilon/3,$$

$x_k \in \tilde{F}_{\epsilon/3}$ and in view of Proposition 2.3, $x_k \in F_\epsilon$. This completes the proof of Theorem 6.6.

6.5   Extension of Theorem 6.6 with Summable Errors

Theorem 6.7 *Assume that* $\Lambda \in (0, 2^{-1})$, $\{\Delta_i\}_{i=0}^{\infty} \subset [0, \infty)$,

$$\Delta = \sum_{i=0}^{\infty} \Delta_i < \infty,$$

$$\|T_i(x) - T_i(y)\| \le \|x - y\|, \ x, y \in C, \ i = 1, \dots, m, \tag{6.111}$$

$$\{(\Omega_i, w_i)\}_{i=0}^{\infty} \subset \mathcal{M}_*,$$

for each integer $i \ge 0$,

$$\alpha_i = (\alpha_{i,1}, \dots, \alpha_{i,m}) \in (\Lambda, 1 - \Lambda)^m, \tag{6.112}$$

for each integer $n \ge 0$,

$$(\Omega_{n+\bar{N}}, w_{n+\bar{N}}) = (\Omega_n, w_n), \ \alpha_{n+\bar{N}} = \alpha_n, \tag{6.113}$$

for each integer $j \ge 0$,

$$\{1, \dots, m\} \subset \cup_{i=j}^{j+\bar{N}-1}(\cup_{t \in \Omega_i}\{t_1, \dots, t_{p(t)}\}), \tag{6.114}$$

$\{x_i\}_{i=0}^{\infty} \subset C$, $\{\lambda_i\}_{i=1}^{\infty} \subset [0, \infty)$,

$$\|x_0\| \le M_0,$$

$$(x_{n+1}, \lambda_{n+1}) \in A(x_n, (\Omega_n, w_n), \alpha_n, \Delta_n), \ n = 0, 1, \dots, \tag{6.115}$$

$$\epsilon \in (0, 1), \ \tilde{\Delta}_0 = \Delta(\bar{m} + 1), \ \epsilon_1 = \epsilon(\bar{N} + 1)^{-1}(9\bar{m} + 9)^{-1}, \tag{6.116}$$

$$\epsilon_0 = 16^{-1}\epsilon_1 \Lambda^2 \widehat{\Delta}(\bar{m}\bar{N})^{-1}\eta(18M_0 + 6\tilde{\Delta}_0 + 1, \epsilon_1(18M_0 + 6\tilde{\Delta}_0 + 1)^{-1}), \tag{6.117}$$

$$N_0 = \lfloor 16M_0\bar{N}\epsilon_0^{-1}\Lambda^{-2}\widehat{\Delta}^{-1}\eta^{-1}(18M_0 + 6\tilde{\Delta}_0 + 1, \epsilon_0(18M_0 + 6\tilde{\Delta}_0 + 1)^{-1})\rfloor, \tag{6.118}$$

$n_1 \ge 0$ *is an integer,*

$$\sum_{i=n_1}^{\infty} \Delta_i < \epsilon_1. \tag{6.119}$$

Then for each integer $n \ge n_1 + (N_0 + 1)\bar{N}$, $x_n \in F_{\epsilon}$.

Proof For each integer $n \geq 0$, each $t = (t_1, \ldots, t_{p(t)}) \in \Omega_n$, and each $i \in \{0, \ldots, p(t)\}$, define $y_{n,t}$, $y_i^{(n,t)}$, $\lambda_{n,t}$ as in Lemma 6.1 such that (6.14)–(6.20) hold. Lemma 6.1 implies that

$$\|x_n\| \leq 3M_0 + \Delta(\bar{m} + 1), \, , \quad n = 0, 1, \ldots. \tag{6.120}$$

Set

$$\tilde{x}_{n_1} = x_{n_1} \tag{6.121}$$

and let $\{\tilde{x}_n\}_{n=n_1+1}^{\infty} \subset C$, $\{\tilde{\lambda}_n\}_{n=n_1+1}^{\infty} \subset [0, \infty)$ be such that for each integer $n \geq n_1$,

$$(\tilde{x}_{n+1}, \tilde{\lambda}_{n+1}) \in A(\tilde{x}_n, (\Omega_n, w_n), \alpha_n, 0). \tag{6.122}$$

Theorem 6.6 and equations (6.111)–(6.113), (6.120), (6.122) imply that for each integer $n \geq n_1 + N_0 \bar{N} + \bar{N}$,

$$\tilde{x}_n \in F_{\epsilon/3}. \tag{6.123}$$

Recall that the sum over an empty set is zero. We show that for each integer $n \geq n_1$,

$$\|x_n - \tilde{x}_n\| \leq \sum \{\Delta_i(\bar{m} + 1) : i \in \{n_1, \ldots, n\} \setminus \{n\}\}. \tag{6.124}$$

In view of (6.121), equation (6.124) holds for $n = n_1$. Assume that $n \geq n_1$ is an integer and (6.124) holds. By (6.122), for each $t = (t_1, \ldots, t_{p(t)}) \in \Omega_n$ and each $i \in \{0, \ldots, p(t)\}$, there exist $\tilde{y}_t \in C$, $\tilde{y}_i^{(t)} \in C$ such that

$$\tilde{y}_0^{(t)} = \tilde{x}_n, \quad \tilde{y}_{p(t)}^{(t)} = \tilde{y}_t, \tag{6.125}$$

for each $i \in \{0, \ldots, p(t) - 1\}$,

$$\tilde{y}_{i+1}^{(t)} = T_{t_{i+1}}, \alpha_{t_{i+1}}(\tilde{y}_i^{(t)}), \tag{6.126}$$

$$\tilde{x}_{n+1} = \sum_{t \in \Omega_n} w_n(t) \tilde{y}_t. \tag{6.127}$$

Let $t = (t_1, \ldots, t_{p(t)}) \in \Omega_n$. In view of (6.19) and (6.125),

$$\|y_0^{(n,t)} - \tilde{y}_0^{(t)}\| = \|x_n - \tilde{x}_n\|. \tag{6.128}$$

We show that for each $i \in \{0, \ldots, p(t)\}$,

$$\|y_i^{(n,t)} - \tilde{y}_i^{(t)}\| \leq \|x_n - \tilde{x}_n\| + \Delta_n i. \tag{6.129}$$

In view of (6.128), relation (6.129) holds for $i = 0$. Assume that $i \in \{0, \ldots, p(t)\} \setminus \{p(t)\}$ and (6.129) holds. It follows from (6.111), (6.119), (6.126), and (6.129) that

$$\|y_{i+1}^{(n,t)} - \tilde{y}_{i+1}^{(t)}\| \leq \|y_{i+1}^{(n,t)} - T_{t_{i+1}}, \alpha_{t_{i+1}}(y_i^{(n,t)})\|$$

$$+ \|T_{t_{i+1}}, \alpha_{t_{i+1}}(y_i^{(n,t)}) - T_{t_{i+1},\alpha_{t_{i+1}}}(\tilde{y}_i^{(t)})\| \leq \Delta_n + \|y_i^{(n,t)} - \tilde{y}_i^{(t)}\|$$

$$\leq \Delta_n + i\Delta_n + \|x_n - \tilde{x}_n\|.$$

Thus we showed by induction that for each $i \in \{0, \ldots, p(t)\}$, (6.129) holds and in view (6.118) and (6.125),

$$\|y_{n,t} - \tilde{y}_t\| \leq \|y_{p(t)}^{(n,t)} - \tilde{y}_{p(t)}^{(t)}\| \leq \|x_n - \tilde{x}_n\| + \Delta_n \bar{m}. \tag{6.130}$$

By (6.16), (6.124), (6.130), and the convexity of the norm,

$$\|x_{n+1} - \tilde{x}_{n+1}\| \leq \|x_{n+1} - \sum_{t \in \Omega_n} w_n(t) y_{n,t}\|$$

$$+ \|\sum_{t \in \Omega_n} w_n(t) y_{n,t} - \sum_{t \in \Omega_n} w_n(t) \tilde{y}_t\|$$

$$\leq \Delta_n + \sum_{t \in \Omega_n} w_n(t) \|y_{n,t} - \tilde{y}_t\|$$

$$\leq \|x_n - \tilde{x}_n\| + \Delta_n(\bar{m} + 1) \leq (\bar{m} + 1) \sum_{i=n_1}^{n} \Delta_i.$$

Thus we have shown by induction that for all integers $n \geq n_1$ relation (6.124) holds, and in view of (6.129),

$$\|x_n - \tilde{x}_n\| \leq (\bar{m} + 1) \sum_{n=n_1}^{\infty} \Delta_n < \epsilon_1(\bar{m} + 1) < \epsilon/9.$$

Let an integer $n \geq n_1 + (N_0 + 1)\bar{N}$. Then $\|x_n - \tilde{x}_n\| < \epsilon/9$ and by (6.123) $\tilde{x}_n \in F_{\epsilon/3}$. Together with (6.111) this implies that for each $s \in \{1, \ldots, m\}$

$$\|x_n - T_s(x_n)\| \leq \|x_n - \tilde{x}_n + \|\tilde{x}_n - T_s(\tilde{x}_n)\| + \|T_x(\tilde{x}_n) - T_s(x_n)\|$$

$$\leq 2\|x_n - x_s\| + \epsilon/3 \leq 2\epsilon/9 + \epsilon/3 < \epsilon.$$

Theorem 6.7 is proved.

6.6   Inexact Iterates with Nonsummable Errors

In this section we prove two theorems which describe the behavior of inexact iterates of our algorithm with nonsummable errors. These results show that if computational errors are small enough, then our method generates approximate solution belonging to the set $\tilde{F}_\gamma$ with $\gamma > 0$. Our first result shows the dependence of γ on our computational errors and calculates a number of iterates which should be done in order to obtain this approximate solution.

Recall that we denote by $\mathrm{Card}(A)$ the cardinality of a set A and suppose that the sum over empty set is zero.

Theorem 6.8 *Let* $\Lambda \in (0, 2^{-1})$, $\epsilon_0 \in (0, 1)$,

$$\epsilon = \epsilon_0(\bar{m} + 1)(2 + 2\bar{N}), \tag{6.131}$$

$$0 < \delta_0 \le 64^{-1}\epsilon_0\widehat{\Delta}\Lambda^2\eta(6M_0 + 3, \epsilon_0(6M_0 + 3)^{-1})(\bar{m}\bar{N})^{-1}, \tag{6.132}$$

$$n_0 = \lfloor 32M_0\epsilon_0^{-1}\widehat{\Delta}^{-1}\Lambda^{-2}\eta^{-1}(6M_0 + 3, \epsilon_0(6M_0 + 3)^{-1})\rfloor + 1. \tag{6.133}$$

Assume that

$$\{(\Omega_i, w_i)\}_{i=1}^\infty \subset \mathcal{M}_*,$$

for each integer $i \ge 0$,

$$\alpha_i = (\alpha_{i,1}, \ldots, \alpha_{i,m}) \in (\Lambda, 1 - \Lambda)^m, \tag{6.134}$$

for each integer $j \ge 0$,

$$\{1, \ldots, m\} \subset \cup_{i=j}^{j+\bar{N}-1}(\cup_{t\in\Omega_i}\{t_1, \ldots, t_{p(t)}\}), \tag{6.135}$$

$$x_0 \in B(0, M_0) \cap C, \tag{6.136}$$

$\{x_i\}_{i=1}^\infty \subset C$, $\{\lambda_i\}_{i=1}^\infty \subset [0, \infty)$, *for each integer* $n \ge 0$,

$$(x_{n+1}, \lambda_{n+1}) \in A(x_n, (\Omega_n, w_n), \alpha_n, \delta_0). \tag{6.137}$$

Then there exists an integer $q \in [0, n_0]$ *such that*

$$\|x_i\| \le 3M_0 + 1, \quad i = 0, \ldots, q\bar{N},$$

$$\lambda_i \le \epsilon_0, \quad i = q\bar{N} + 1, \ldots, (q + 1)\bar{N}. \tag{6.138}$$

Moreover, if an integer $q \in [0, n_0]$ satisfies (6.138), then for each $i = q\bar{N}, \ldots, (q + 1)\bar{N}$,

$$x_i \in \tilde{F}_{\epsilon_1}$$

and

$$\|x_i - x_j\| \le \epsilon$$

for each $i, j \in \{q\bar{N}, \ldots, (q + 1)\bar{N}\}$.

Proof Set

$$\Delta_n = \delta_0, \quad n = 0, 1, \ldots. \tag{6.139}$$

Let for each integer $n \ge 0$, each $t = (t_1, \ldots, t_{p(t)}) \in \Omega_n$, and each $i \in \{0, \ldots, p(t)\}$, $y_{n,t}$, $y_i^{(n,t)}$, $\lambda_{n,t}$ be defined as in Lemma 6.1 such that (6.14)–(6.20) hold with $\Delta_n = \delta_0$, $n = 0, 1, \ldots$. Assume that a nonnegative integer s satisfies for each integer $k \in [0, s]$,

$$\max\{\lambda_i : i = k\bar{N} + 1, \ldots, (k + 1)\bar{N}\} > \epsilon_0. \tag{6.140}$$

Choose

$$\delta \in (0, 2^{-1}\delta_0). \tag{6.141}$$

By assumption (A) and (6.136),

$$\|x_0 - p_\delta\| \le 2M_0. \tag{6.142}$$

Assume that an integer $k \in [0, s]$ satisfies

$$\|x_{k\bar{N}} - z\| \le 2M_0. \tag{6.143}$$

We prove the following auxiliary result.

Lemma 6.9 *Assume that an integer $i \in [0, \bar{N} - 1]$ satisfies*

$$\|x_{k\bar{N}+i} - p_\delta\| \le 2M + i(\bar{m} + 1)(\delta + \delta_0).$$

Then

$$\|x_{k\bar{N}+i+1} - p_\delta\| \le \|x_{k\bar{N}+i} - p_\delta\| + (\bar{m}+1)(\delta + \delta_0), \tag{6.144}$$

$$\|x_{k\bar{N}+i+1} - p_\delta\| \le 2M + (\delta + \delta_0)(\bar{m}+1)(i+1). \tag{6.145}$$

If $\lambda_{k\bar{N}+i+1} > \epsilon_0$, *then*

$$\|x_{k\bar{N}+i+1} - p_\delta\| \le \|x_{k\bar{N}+i} - p_\delta\| - 8^{-1}\widehat{\Delta\epsilon}_0\Lambda^2\eta(6M_0+3, \epsilon_0(6M_0+3)^{-1}). \tag{6.146}$$

Proof Clearly, (6.145) follows from (6.144). Assertion 1 of Lemma 6.1 and assumption (A) imply that for each $t = (t_1, \ldots, t_{p(t)}) \in \Omega_{k\bar{N}+i}$ and each $j \in \{0, \ldots, p(t)\}$,

$$\|y_j^{(k\bar{N}+i,t)} - p_\delta\| \le \|x_{k\bar{N}+i} - p_\delta\| + \bar{m}(\delta + \delta_0)\bar{m} + \delta \le 2M_0 + 1, \tag{6.147}$$

$$\|y_j^{(k\bar{N}+i,t)}\| \le 3M_0 + 1. \tag{6.148}$$

Assume that

$$\lambda_{k\bar{N}+i+1} > \epsilon_0. \tag{6.149}$$

In view of (6.14), (6.17), and (6.149), there exists an index vector

$$s = (s_1, \ldots, s_{p(s)}) \in \Omega_{k\bar{N}+i+1}$$

such that

$$\lambda_{k\bar{N}+i,s} = \lambda_{k\bar{N}+i+1} > \epsilon_0,$$

and there exists an integer

$$j_0 \in \{0, \ldots, p(s) - 1\}$$

such that

$$\|y_{j_0}^{(k\bar{N}+i,s)} - T_{s_{j_0+1}}(y_{j_0}^{(k\bar{N}+i,s)})\| = \lambda_{k\bar{N}+i,s} > \epsilon_0. \tag{6.150}$$

By assumption (A), equations (6.132), (6.134), (6.141), (6.148), (6.150), and Lemma 2.4 applied with $S = T_{s_{j_0+1}}$, $u = y_{j_0}^{(k\bar{N}+i,s)}$, $\alpha = \alpha_{s_{j_0+1}}$, $M = 3M_0 + 1$, $p = p_\delta$, $\gamma = \epsilon_0$, and

$$v = (1 - \alpha_{s_{j_0+1}})y_{j_0}^{(k\bar{N}+i,s)} + \alpha_{s_{j_0+1}}T_{s_{j_0+1}}(y_{j_0}^{(k\bar{N}+i,s)}),$$

we have

$$\|p_\delta - (1 - \alpha_{s_{j_0+1}})y_{j_0}^{(k\bar{N}+i,s)} + \alpha_{s_{j_0+1}}T_{s_{j_0+1}}(y_{j_0}^{(k\bar{N}+i,s)})\|$$

$$\leq \|y_{j_0}^{(k\bar{N}+i,s)} - p_\delta\| - 4^{-1}\epsilon_0\Lambda^2\eta(6M_0 + 3, \epsilon_0(6M_0 + 3)^{-1}).$$

Together with (6.19) and Lemma 6.1 with $\Delta_n = \delta_0, n = 0, 1, \ldots$, this implies that

$$\|y_{j_0+1}^{(k\bar{N}+i,s)} - p_\delta\| \leq \delta_0 + \|y_{j_0}^{(k\bar{N}+i,s)} - p_\delta\|$$

$$- 4^{-1}\epsilon_0\Lambda^2\eta(6M_0 + 3, \epsilon_0(6M_0 + 3)^{-1}). \tag{6.151}$$

Assertion 1 of Lemma 6.1 (with $\Delta_n = \delta_0, \ n = 0, 1, \ldots$) and equations (6.18), (6.151) imply that

$$\|x_{k\bar{N}+i} - p_\delta\| - \|y_{k\bar{N}+i,s} - p_\delta\|$$

$$= \|y_0^{(k\bar{N}+i,s)} - p_\delta\| - \|y_{p(s)}^{(k\bar{N}+i,s)} - p_\delta\|$$

$$= \sum_{j=0}^{p(t)-1}(\|y_j^{(k\bar{N}+i,s)} - p_\delta\| - \|y_{j+1}^{(k\bar{N}+i,s)} - p_\delta\|)$$

$$\geq 4^{-1}\epsilon_0\Lambda^2\eta(6M_0 + 3, \epsilon_0(6M_0 + 3)^{-1}) - \delta_0 - (\delta + \delta_0)(p(s) - 1). \tag{6.152}$$

In view of (6.147), for each $t \in \Omega_{k\bar{N}+i}$,

$$\|y_{k\bar{N}+i,t} - p_\delta\| \leq \|x_{k\bar{N}+i} - p_\delta\| + (\delta + \delta_0)(\bar{m} + 1). \tag{6.153}$$

By (6.14)–(6.16) (with $\Delta_n = \delta_0$), (6.132), (6.137), (6.141), (6.152), and the convexity of the norm,

$$\|x_{k\bar{N}+i+1} - p_\delta\| \leq \|x_{k\bar{N}+i+1} - \sum_{t\in\Omega_{k\bar{N}+i}} w_{k\bar{N}+i}(t)y_{k\bar{N}+i,t}\|$$

$$+ \|\sum_{t\in\Omega_{k\bar{N}+i}} w_{k\bar{N}+i}(t)y_{k\bar{N}+i,t} - p_\delta\|$$

$$\leq \delta_0 + \sum_{t\in\Omega_{k\bar{N}+i}} w_{k\bar{N}+i}(t)\|y_{k\bar{N}+i,t} - p_\delta\|$$

$$\leq \delta_0 + w_{k\bar{N}+i}(s)\|y_{k\bar{N}+i,s} - p_\delta\|$$

$$+ \sum \{w_{k\bar{N}+i}(t) \|y_{k\bar{N}+i,t} - p_\delta\| : t \in \Omega_{k\bar{N}+i} \setminus \{s\}\}$$

$$\leq \delta_0 + w_{k\bar{N}+i}(s)(\|x_{k\bar{N}+i} - p_\delta\| + \delta_0 + (\delta + \delta_0)(p(s) - 1)$$

$$- 4^{-1} \epsilon_0 \Lambda^2 \eta(6M_0 + 3, \epsilon_0(6M_0 + 3)^{-1})$$

$$+ (1 - w_{k\bar{N}+i}(s))(\|x_{k\bar{N}+i} - p_\delta\| + (\bar{m} + 1)(\delta + \delta_0))$$

$$\leq \|x_{k\bar{N}+i} - p_\delta\| + (\bar{m} + 1)(\delta + \delta_0) + \delta_0 - 4^{-1} \epsilon_0 \widehat{\Delta} \Lambda^2 \eta(6M_0 + 3, \epsilon_0(6M_0 + 3)^{-1})$$

$$\leq \|x_{k\bar{N}+i} - p_\delta\| - 8^{-1} \epsilon_0 \widehat{\Delta} \Lambda^2 \eta(6M_0 + 3, \epsilon_0(6M_0 + 3)^{-1}).$$

Lemma 6.9 is proved.

It follows from (6.132), (6.140), (6.141), (6.143), and Lemma 6.9 applied by induction that for all integers $i = 0, \ldots, \bar{N} - 1$,

$$\|x_{k\bar{N}+i+1} - p_\delta\| \leq \|x_{k\bar{N}+i} - p_\delta\| + (\delta + \delta_0)(\bar{m} + 1), \tag{6.154}$$

$$\|x_{k\bar{N}+i+1} - p_\delta\| \leq 2M + (\delta + \delta_0)(\bar{m} + 1)\bar{N} \leq 2M_0 + 1. \tag{6.155}$$

Lemma 6.9 and (6.132), (6.141), and (6.154) imply that

$$\|x_{(k+1)\bar{N}} - p_\delta\| - \|x_{k\bar{N}} - p_\delta\|$$

$$= \sum_{i=0}^{\bar{N}-1} [\|x_{k\bar{N}+i+1} - p_\delta\| - \|x_{k\bar{N}+i} - p_\delta\|]$$

$$\leq -8^{-1} \widehat{\Delta} \epsilon_0 \Lambda^2 \eta(6M_0 + 3, \epsilon_0(6M_0 + 3)^{-1}) + \bar{N}(\delta + \delta_0)(\bar{m} + 1)$$

$$\leq -16^{-1} \widehat{\Delta} \epsilon_0 \Lambda^2 \eta(6M_0 + 3, \epsilon_0(6M_0 + 3)^{-1}).$$

Thus we have shown that the following property holds:

(a) if an integer $k \in [0, s]$ satisfies $\|x_{k\bar{N}} - p_\delta\| \leq 2M_0$, then

$$\|x_j - p_\delta\| \leq 2M_0 + 1, \quad j = k\bar{N}, \ldots, (k + 1)\bar{N},$$

$$\|x_{(k+1)\bar{N}} - p_\delta\| \leq \|x_{k\bar{N}} - p_\delta\| - 16^{-1} \widehat{\Delta} \epsilon_0 \Lambda^2 \eta(6M_0 + 3, \epsilon_0(6M_0 + 3)^{-1}).$$

In view of (6.142) and property (a), we have that

$$\|x_j - p_\delta\| \leq 2M_0 + 1, \quad j = 0, \ldots, (s + 1)\bar{N},$$

(6.155) is true for every integer $k = 0, \ldots, s$ and that

$$16^{-1}(s+1)\widehat{\Delta}\epsilon_0 \Lambda^2 \eta(6M_0 + 3, \epsilon_0(6M_0 + 3)^{-1})$$

$$\leq \sum_{k=0}^{s}(\|x_{k\bar{N}} - p_\delta\| - \|x_{(k+1)\bar{N}} - p_\delta\|) \leq \|x_0 - p_\delta\| \leq 2M_0,$$

$$s+1 \leq 32M_0\widehat{\Delta}^{-1}\Lambda^{-2}\epsilon_0^{-1}\eta^{-1}(6M_0 + 3, \epsilon_0(6M_0 + 3)^{-1}).$$

Thus we have shown that the following property holds:

(b) If for an integer $s \geq 0$ and for every integer $k \in [0, s]$ relation (6.140) holds, then

$$s \leq 32M_0\widehat{\Delta}^{-1}\Lambda^{-2}\epsilon_0^{-1}\eta^{-1}(6M_0 + 3, \epsilon_0(6M_0 + 3)^{-1}) - 1,$$

$$\|x_j - p_\delta\| \leq 2M_0 + 1, \ \ j = 0, \ldots, (s+1)\bar{N},$$

$$\|x_{k\bar{N}} - p_\delta\| \leq 2M_0, \ \ k = 0, \ldots, s+1.$$

Property (b) implies that there exists an integer $q \in [0, n_0]$ such that for each integer k satisfying $0 \leq k < q$,

$$\max\{\lambda_i : \ i = k\bar{N} + 1, \ldots, (k+1)\bar{N}\} > \epsilon_0,$$

$$\max\{\lambda_i : \ i = q\bar{N} + 1, \ldots, (q+1)\bar{N}\} \leq \epsilon_0, \tag{6.156}$$

$$\|x_{q\bar{N}} - p_\delta\| \leq 2M_0,$$

$$\|x_j - p_\delta\| \leq 2M_0 + 1, \ \ j = 0, \ldots, q\bar{N},$$

$$\|x_{q\bar{N}}\| \leq 3M_0, \ \|x_j\| \leq 3M_0 + 1, \ \ j = 0, \ldots, q\bar{N}.$$

Assume that an integer $q \in [0, n_0]$ satisfies (6.156). Then for each $i \in \{q\bar{N} + 1, \ldots, (q+1)\bar{N}\}$,

$$\lambda_i \leq \epsilon_0. \tag{6.157}$$

Let

$$k \in \{q\bar{N}, \ldots, (q+1)\bar{N} - 1\}.$$

It follows from (6.17), (6.20), and (6.157) that for every index vector $t = (t_1, \ldots, t_{p(t)}) \in \Omega_k$, for every integer $j \in \{0, \ldots, p(t) - 1\}$,

$$\|y_j^{(k,t)} - T_{t_{j+1}}(y_j^{(k,t)})\| \leq \epsilon_0, \tag{6.158}$$

in view of (6.19) with $\Delta_k = \delta_0$ and (6.18),

$$\|y_j^{(k,t)} - y_{j+1}^{(k,t)}\| \leq \epsilon_0 + \delta_0,$$

$$\|x_k - y_{j+1}^{(k,t)}\| \leq (\epsilon_0 + \delta_0)\bar{m}. \tag{6.159}$$

Together with (6.18) (with $\Delta_k = \delta_0$) and the convexity of the norm, this implies that

$$\|x_k - y_{k,t}\| \leq (\epsilon_0 + \delta_0)\bar{m}, \ t \in \Omega_k,$$

$$\|x_{k+1} - x_k\| \leq \|x_{k+1} - \sum_{t \in \Omega_k} w_k(t)y_{k,t}\|$$

$$+ \|\sum_{t \in \Omega_k} w_k(t)y_{k,t} - x_k\|$$

$$\leq \delta_0 + \sum_{t \in \Omega_k} w_k(t)\|y_{k,t} - x_k\| \leq (\delta_0 + \epsilon_0)(\bar{m} + 1). \tag{6.160}$$

In view of (6.160), for each $k_1, k_1 \in \{q\bar{N}, \ldots, (q+1)\bar{N}\}$,

$$\|x_{k_1} - x_{k_2}\| \leq (\delta_0 + \epsilon_0)(\bar{m} + 1)\bar{N}. \tag{6.161}$$

Let

$$l \in \{q\bar{N}, \ldots, (q+1)\bar{N}\}.$$

It follows from (6.131), (6.132), (6.159), and (6.161) that for each $k \in \{q\bar{N}, \ldots, (q+1)\bar{N} - 1\}$, each $t = (t_1, \ldots, t_{p(t)}) \in \Omega_k$, and every integer $j \in \{0, \ldots, p(t)\}$,

$$\|x_l - y_j^{(k,t)}\| \leq \|x_l - x_k\| + \|x_k - y_j^{(k,t)}\|$$

$$\leq (\delta_0 + \epsilon_0)(\bar{m} + 1)\bar{N} + \epsilon_0\bar{m} + \delta_0\bar{m} < (\bar{m} + 1)(\bar{N} + 1)(\delta_0 + \epsilon_0) = \epsilon. \tag{6.162}$$

Let $s \in \{1, \ldots, m\}$. By (6.135), there exist an integer $k \in \{q\bar{N}, \ldots, (q+1)\bar{N} - 1\}$, an index vector $t = (t_1, \ldots, t_{p(t)}) \in \Omega_k$ and $j \in \{0, \ldots, p(t) - 1\}$ such that

$$s = t_{j+1}.$$

Together with (6.158), this implies that

$$\| y_j^{(k,t)} - T_s(y_j^{(k,t)}) \| \leq \epsilon_0.$$

Combined with (6.162), this implies that $x_l \in \tilde{F}_\epsilon$ and completes the proof of Theorem 6.8.

Theorem 6.8 implies the following result.

Theorem 6.10 *Let us assume that* $\Lambda \in (0, 2^{-1})$, $\bar{\epsilon} \in (0, 1)$, $\tilde{F}_{\bar{\epsilon}} \subset B(0, M_0)$,

$$\epsilon \in (0, \bar{\epsilon}], \quad \epsilon_0 = \epsilon(\bar{m} + 1)^{-1}(2 + 2\bar{N}),$$

$$0 < \delta_0 < 64^{-1}\epsilon_0 \hat{\Delta} \Lambda^2 \eta(6M_0 + 3, \epsilon_0(6M_0 + 3)^{-1})(\bar{m}\bar{N})^{-1},$$

$$n_0 = \lfloor 32M_0\epsilon_0^{-1} \hat{\Delta}^{-1} \Lambda^{-2} \eta^{-1}(6M_0 + 3, \epsilon_0(6M_0 + 3)^{-1}) \rfloor + 1.$$

Assume that $\{(\Omega_i, w_i)\}_{i=1}^{\infty} \subset \mathcal{M}_*$, *for each integer* $i \geq 0$,

$$\alpha_i = (\alpha_{i,1}, \ldots, \alpha_{i,m}) \in (\Lambda, 1 - \Lambda)^m,$$

for each integer $j \geq 0$,

$$\{1, \ldots, m\} \subset \cup_{i=j}^{j+\bar{N}-1}(\cup_{t \in \Omega_i}\{t_1, \ldots, t_{p(t)}\}),$$

$$x_0 \in B(0, M_0) \cap C,$$

$\{x_i\}_{i=1}^{\infty} \subset C$, $\{\lambda_i\}_{i=1}^{\infty} \subset [0, \infty)$, *for each integer* $n \geq 0$,

$$(x_{n+1}, \lambda_{n+1}) \in A(x_n, (\Omega_n, w_n), \alpha_n, \delta_0).$$

Then

$$\|x_i\| \leq 3M_0 + 1, \quad i = 0, 1, \ldots,$$

and there exists a strictly increasing sequence of integers $\{q_k\}_{k=0}^{\infty}$ *such that*

$$0 \leq q_0 \leq n_0$$

and that for each integer $k \geq 0$ and each $i \in \{q_k \bar{N}, \ldots, (q_k+1)\bar{N}\}$, $1 \leq q_{k+1} - q_k \leq n_0$ and $x_i \in F_\epsilon$.

Theorem 6.11 *Let $\Lambda \in (0, 2^{-1})$, $\bar{\epsilon} \in (0, 1)$,*

$$\tilde{F}_{\bar{\epsilon}} \subset B(0, M_0), \tag{6.163}$$

$$\|T_i(x) - T_i(y)\| \leq \|x - y\|, \ x, y \in C, \ i = 1, \ldots, m, \tag{6.164}$$

$$\epsilon \in (0, \bar{\epsilon}), \epsilon_1 = \epsilon \bar{m}^{-1}(18 + 18\bar{N})^{-1}, \tag{6.165}$$

$$\epsilon_0 = 16^{-1}\epsilon_1 \widehat{\Delta} \eta (6M_0 + 1, \epsilon_1(6M_0 + 1)^{-1})(\bar{m}\bar{N})^{-1}, \tag{6.166}$$

$$N_0 = \lfloor 16M_0\bar{N}\epsilon_0^{-1}\widehat{\Delta}^{-1}\Lambda^{-2}\eta^{-1}(6M_0 + 1, \epsilon_0(6M_0 + 1)^{-1})\rfloor + 1, \tag{6.167}$$

$$\delta = (8N_0\bar{N}(\bar{m} + 1))^{-1}\epsilon. \tag{6.168}$$

Assume that $\{(\Omega_i, w_i)\}_{i=1}^{\infty} \subset \mathcal{M}_*$, for each integer $i \geq 0$,

$$\alpha_i = (\alpha_{i,1}, \ldots, \alpha_{i,m}) \in (\Lambda, 1 - \Lambda)^m, \tag{6.169}$$

for each integer $n \geq 0$,

$$(\Omega_{n+\bar{N}}, w_{n+\bar{N}}) = (\Omega_n, w_n), \ \alpha_{n+\bar{N}} = \alpha_n, \tag{6.170}$$

for each integer $j \geq 0$,

$$\{1, \ldots, m\} \subset \cup_{i=j}^{j+\bar{N}-1}(\cup_{t \in \Omega_i}\{t_1, \ldots, t_{p(t)}\}), \tag{6.171}$$

$$x_0 \in B(0, M_0) \cap C, \tag{6.172}$$

$\{x_i\}_{i=1}^{\infty} \subset C$, $\{\lambda_i\}_{i=1}^{\infty} \subset [0, \infty)$, for each integer $n \geq 0$,

$$(x_{n+1}, \lambda_{n+1}) \in A(x_n, (\Omega_n, w_n), \alpha_n, \delta). \tag{6.173}$$

Then for each integer $n \geq N_0\bar{N}$, $x_n \in F_\epsilon$.

Proof For each integer $n \geq 0$, each $t = (t_1, \ldots, t_{p(t)}) \in \Omega_n$, and each $i \in \{0, \ldots, p(t)\}$, define $y_{n,t}$, $y_i^{(n,t)}$, $\lambda_{n,t}$ as in Lemma 6.1 such that (6.14)–(6.20) hold with $\Delta_j = \delta$, $j = 0, 1, \ldots$.

Assume that $s \geq 0$ is an integer and

$$\|x_s\| \leq M_0. \tag{6.174}$$

Set

$$\tilde{x}_s = x_s \tag{6.175}$$

and let $\{\tilde{x}_n\}_{n=s+1}^{\infty} \subset C, \{\mu_n\}_{n=s+1}^{\infty} \subset [0, \infty)$ be such that for each integer $n \geq s$,

$$(\tilde{x}_{n+1}, \mu_{n+1}) \in A(\tilde{x}_n, (\Omega_n, w_n), \alpha_n, 0). \tag{6.176}$$

Theorem 6.6 and equations (6.169)–(6.176) imply that for each integer $n \geq s + N_0\bar{N} + \bar{N}$,

$$\tilde{x}_n \in F_{\epsilon/2}. \tag{6.177}$$

We show that for each integer $n \geq s$,

$$\|x_n - \tilde{x}_n\| \leq (n - s)\delta(\bar{m} + 1). \tag{6.178}$$

In view of (6.175), equation (6.178) holds for $n = s$. Assume that $n \geq s$ is an integer and (6.178) holds. By (6.176), for each $t = (t_1, \ldots, t_{p(t)}) \in \Omega_n$ and each $i \in \{0, \ldots, p(t)\}$, there exist $y_t \in C, y_i^{(t)} \in C$ such that

$$y_0^{(t)} = \tilde{x}_n, \tag{6.179}$$

for each $i \in \{0, \ldots, p(t) - 1\}$,

$$y_{i+1}^{(t)} = T_{t_{i+1}}, \alpha_{t_{i+1}}(y_i^{(t)}), \tag{6.180}$$

$$y_{p(t)}^{(t)} = y_t, \tag{6.181}$$

$$\tilde{x}_{n+1} = \sum_{t \in \Omega_n} w_n(t) y_t. \tag{6.182}$$

Let $t = (t_1, \ldots, t_{p(t)}) \in \Omega_n$ and $i \in \{0, \ldots, p(t) - 1\}$. In view of (6.19), (6.164), and (6.174),

$$\|y_{i+1}^{(n,t)} - y_{i+1}^{(t)}\| \leq \|y_{i+1}^{(n,t)} - T_{t_{i+1}}, \alpha_{t_{i+1}}(y_i^{(n,t)})\|$$

$$+ \|T_{t_{i+1}}, \alpha_{t_{i+1}}(y_i^{(n,t)}) - T_{t_{i+1}, \alpha_{t_{i+1}}}(y_i^{(t)})\| \leq \delta + \|y_i^{(n,t)} - y_i^{(t)}\|$$

and

$$\|y_{n,t} - \tilde{y}_t\| = \|y_{p(t)}^{(n,t)} - y_{p(t)}^{(t)}\| \leq \|\tilde{x}_n - x_n\| + \bar{m}\delta. \tag{6.183}$$

By (6.16), (6.183), and the convexity of the norm,

$$\|x_{n+1} - \tilde{x}_{n+1}\| \leq \|x_{n+1} - \sum_{t \in \Omega_n} w_n(t) y_{n,t}\|$$

$$+ \|\sum_{t \in \Omega_n} w_n(t) y_{n,t} - \sum_{t \in \Omega_n} w_n(t) \tilde{y}_t\|$$

$$\leq \delta + \sum_{t \in \Omega_n} w_n(t) \|y_{n,t} - \tilde{y}_t\|$$

$$\leq \|x_n - \tilde{x}_n\| + \delta(\bar{m} + 1).$$

Thus we have shown by induction that for all integers $n \geq s$ relation (6.178) holds.
Let

$$n \in \{s + \bar{N}(N_0 + 1), \ldots, s + \bar{N}(2N_0 + 1)\}. \tag{6.184}$$

In view of (6.178) and (6.184),

$$\|x_n - \tilde{x}_n\| \leq 2(N_0 + 1)\delta\bar{N}(\bar{m} + 1). \tag{6.185}$$

By (6.164), (6.168), and (6.177), for each $s \in \{1, \ldots, m\}$,

$$\|x_n - T_s(x_n)\| \leq \|x_n - \tilde{x}_n\| + \|\tilde{x}_n - T_s(\tilde{x}_n)\| + \|T_x(\tilde{x}_n) - T_s(x_n)\|$$

$$\leq 2\|x_n - \tilde{x}_n\| + \epsilon/2 \leq 2(N_0 + 1)\bar{N}(\bar{m} + 1)\delta + \epsilon/2 \leq \epsilon$$

and $x_n \in F_\epsilon$. Thus we showed that the following property holds:

(P) Is $s \geq 0$ is an integer and $\|x_s\| \leq M_0$, then $x_n \in F_\epsilon$ for each $n \in \{s + \bar{N}(N_0 + 1), \ldots, s + \bar{N}(2N_0 + 1)\}$.

Property (P) and (6.163), (6.172) imply that $x_n \in F_\epsilon$ for each integer $n \geq \bar{N}(N_0 + 1)$. Theorem 6.11 is proved.

References

1. Agarwal RP, Karapinar Erdal, O'Regan D, Roldán-López-de-Hierro AF (2015) Fixed point theory in metric type spaces. Springer, Cham
2. Alber YI, Li JL (2007) The connection between the metric and generalized projection operators in Banach spaces. Acta Math Sin (Engl Ser) 23:1109–1120
3. Alber YI, Yao JC (2009) Another version of the proximal point algorithm in a Banach space. Nonlinear Anal 70:3159–3171
4. Ariza-Ruiz D, Leustean L, and G. Lopez—Acedo (2024) Firmly nonexpansive mappings in classes of geodesic spaces. Trans Am Math Soc 366:4299–4322
5. Bacak M (2012) Proximal point algorithm in metric spaces. Israel J Math 160:1–13
6. Bacak M (2014) Convex analysis and optimization in Hadamard spaces. De Gruyter series in nonlinear analysis and applications
7. Banach S (1922) Sur les opérations dans les ensembles abstraits et leur application aux équations intégrales. Fund Math 3:133–181
8. Bargetz C, Medjic E (2020) On the rate of convergence of iterated Bregman projections and of the alternating algorithm. J Math Anal Appl 481:123482
9. Bargetz C, Reich S, Thimm D (2023) Generic properties of nonexpansive mappings on unbounded domains. J Math Anal Appl 526:127179
10. Bauschke HH, Combettes PL (2017) Convex analysis and monotone operator theory in Hilbert spaces. CMS Books in Mathematics/Ouvrages de Mathématiques de la SMC. Springer, Cham
11. Blum E, Oettli W (1994) From optimization and variational inequalities to equilibrium problems. Math Student 63:123–145
12. Borwein J, Reich S, Shafrir I (1992) Krasnoselski–Mann iterations in normed spaces. Can. Math Bull 35:21–28
13. Butnariu D, Reich S, Zaslavski AJ (2006) Convergence to fixed points of inexact orbits of Bregman-monotone and of nonexpansive operators in Banach spaces. Fixed point theory and its applications. Yokohama Publishers, Mexico, pp 11–32
14. Butnariu D, Resmerita E (2001) The outer Bregman projection method for stochastic feasibility problems in Banach spaces. Studies in computational mathematics, vol 8. North-Holland Publishing, Amsterdam, pp 69–86.
15. Cegielski A, Gibali A, Reich S, Zalas R (2013) An algorithm for solving the variational inequality problem over the fixed point set of a quasi-nonexpansive operator in Euclidean space. Numer Funct Anal Optim 34:1067–1096

16. Ceng LC, Ansari QH, Perrusel A, Yao JC (2015) Approximation methods for triple hierarchical variational inequalities. Fixed Point Theory 16:67–90
17. Ceng LC, Hadjisavvas N, Wong NC (2010) Strong convergence theorem by a hybrid extragradient-like approximation method for variational inequalities and fixed point problems. J Global Optim 46:635–646
18. Ceng LC, Mordukhovich BS, Yao JC (2010) Hybrid approximate proximal method with auxiliary variational inequality for vector optimization. J Optim Theory Appl 146:267–303
19. Ceng LC, Petrusel A, Qin X, Yao J-C (2021) Pseudomonotone variational inequalities and fixed points. Fixed Point Theory 22:543–558
20. Ceng LC, Plubtieng S, Wong MM, Yao JC (2015) System of variational inequalities with constraints of mixed equilibria, variational inequalities, and convex minimization and fixed point problems. J Nonlinear Convex Anal 16:385–421
21. Ceng LC, Wong NC, Yao JC (2014) Regularized hybrid iterative algorithms for triple hierarchical variational inequalities. J Inequalities Appl 2014:490
22. Ceng LC, Wong NC, Yao JC (2015) Hybrid extragradient methods for fiinding minimum norm solutions of split feasibility problems. J Nonlinear Convex Anal 16:1965–1983
23. Censor Y, Davidi, R, Herman GT (2010) Perturbation resilience and superiorization of iterative algorithms. Inverse Probl 26:1–12
24. Censor Y, Elfving T, Herman GT (2001) Averaging strings of sequential iterations for convex feasibility problems. In: Butnariu D, Censor Y, Reich S (eds) Inherently parallel algorithms in feasibility and optimization and their applications. North-Holland, Amsterdam, pp 101–113
25. Censor Y, Elfving T, Herman GT, Nikazad T (2008) Diagonally-relaxed orthogonal projection methods. SIAM J Sci Comput 30:473–504
26. Censor Y, Lent A (1982) Cyclic subgradient projections. Math. Program 24:233–235
27. Censor Y, Segal A (2009) On the string averaging method for sparse common fixed point problems. Inter Trans Oper Res 16:481–494
28. Censor Y, Segal A (2009) The split common fixed point problem for directed operators. J Convex Anal 16:587–600
29. Censor Y, Zaslavski AJ (2013) Convergence and perturbation resilience of dynamic string-averaging projection methods. Comput Optim Appl 54:65–76
30. Censor Y, Zaslavski AJ (2015) Strict Fejer monotonicity by superiorization of feasibility-seeking projection methods. J Optim Theory Appl 165:172–187
31. Censor Y, Zenios S (1997) Parallel optimization: theory, algorithms and applications. Oxford University Press, New York
32. Cheval H, Kohlenbach U, Leustean L (2023) On modified Halpern and Tikhonov-Mann iterations. J Optim Theory Appl 197:233–251
33. Cheval H, Leustean L (2022) Quadratic rates of asymptotic regularity for the Tikhonov–Mann iteration. Optim Methods Softw 37:2225–2240
34. Cheval H, Leustean L. Linear rates of asymptotic regularity for Halpern-type iterations, arXiv:2303.05406v2 [math.OC]. https://doi.org/10.48550/arXiv.2303.05406
35. Colao V, Leustean L, Lopez-Acedo G, Martin-Marquez V (2011) Alternative iterative methods for nonexpansive mappings, rates of convergence and applications. J Convex Anal 18:465–487
36. Combettes PL (1996) The convex feasibility problem in image recovery. Adv Imaging Electron Phys 95:155–270
37. Combettes PL (1997) Hilbertian convex feasibility problems: convergence of projection methods. Appl Math Optim 35:311–330
38. Combettes PL, Hirstoaga SA (2005) Equilibrium problems in Hilbert spaces. J Nonlinear Convex Anal 63:117–136
39. Dinis B, Pinto P (2023) Strong convergence for the alternating Halpern-Mann iteration in CAT(0) spaces. SIOPT 33:785–815
40. Djafari-Rouhani B, Farid M, Kazmi KR (2016) Common solution to generalized mixed equilibrium problem and fixed point problem for a nonexpansive semigroup in Hilbert space. J Korean Math Soc 53:89–114

41. Djafari-Rouhani B, Kazmi KR, Farid M, (2017) Common solutions to some systems of variational inequalities and fixed point problems. Fixed Point Theory 18:167–190
42. Djafari-Rouhani B, Kazmi KR, Moradi S, Ali R, Khan SA (2022) Solving the split equality hierarchical fixed point problem. Fixed Point Theory 23:351–369
43. Djafari-Rouhani B, Mohebbi V (2020) Proximal point method for quasi-equilibrium problems in Banach spaces. Numer Funct Anal Optim 41:1007–1026
44. Dong Q-L, Cho YJ, He S, Pardalos PM, Themistocles M (2022) The Krasnosel'skii-Mann iterative method–recent progress and applications. SpringerBriefs in optimization. Springer, Cham
45. Du WS (2019) Some generalizations of fixed point theorems of Caristi type and Mizoguchi–Takahashi type under relaxed conditions. Bull Braz Math Soc (NS) 50:603–624
46. Espínola R, Wiśnicki A (2018) The Knaster-Tarski theorem versus monotone nonexpansive mappings. Bull Pol Acad Sci Math 66:1–7
47. Gerhardy P, Kohlenbach U (2008) General logical metatheorems for functional analysis. Trans Am Math Soc 360:2615–2660
48. Gibali A (2017) A new split inverse problem and an application to least intensity feasible solutions. Pure Appl Funct Anal 2:243–258
49. Gibali A, Reich S, Zalas R (2015) Iterative methods for solving variational inequalities in Euclidean space. J Fixed Point Theory Appl 17:775–811
50. Gibali A, Reich S, Zalas R (2017) Outer approximation methods for solving variational inequalities in Hilbert space. Optimization 66:417–437
51. Goebel K, Kirk WA (1990) Topics in metric fixed point theory. Cambridge University Press, Cambridge
52. Goebel K, Reich S (1984) Uniform convexity, hyperbolic geometry, and nonexpansive mappings. Marcel Dekker, New York and Basel
53. Gwinner J (1978) On the convergence of some iteration processes in uniformly convex Banach spaces. Proc Am Math Soc 71:29–35
54. Gwinner J, Jadamba B, Khan AA, Sama M (2018) Identification in variational and quasi-variational inequalities. J Convex Anal 25:545–569
55. Gwinner J, Raciti F (2009) On monotone variational inequalities with random data. J Math Inequal 3:443–453
56. He H, Ling C, Xu, HK (2015) A relaxed projection method for split variational inequalities. J Optim Theory Appl 166:213–233
57. He H, Ling C, Xu, HK (2015) A projection-based splitting method for structured variational inequalities. J Nonlinear Convex Anal 16:1539–1556
58. Iusem A, Resmerita E (2010) A proximal point method in nonreflexive Banach spaces. Set-Valued Var Anal 18:109–120
59. Jachymski J (2008) The contraction principle for mappings on a metric space with a graph. Proc Am Math Soc 136:1359–1373
60. Jadamba B, Khan AA, Sama M (2017) Generalized solutions of quasi-variational inequalities. Optim Lett 6:1221–1231
61. Karapinar E, RP Agarwal (2022) Fixed point theory in generalized metric spaces. Synthesis lectures on mathematics and statistics. Springer, Cham
62. Karapinar E, Mitrovic Z, Ozturk A, Radenovic S (2021) On a theorem of Ciric in b-metric spaces. Rend Circ Mat Palermo 70:217–225
63. Khamsi MA, Kirk WA (2001) An introduction to metric spaces and fixed point theory. Pure and applied thematics (New York). Wiley-Interscience, New York
64. Khamsi MA, Kozlowski W M (2015) Fixed point theory in modular function spaces. Birkhäuser/Springer, Cham
65. Khan AA, Li J (2024) Characterizations of the metric and generalized metric projections on subspaces of Banach spaces. J Math Anal Appl 531(2):127865
66. Khan AA, Li J, Reich S (2023) Generalized projections on general Banach spaces. J Nonlinear Convex Anal 24:1079–1112

67. Khan AA, Tammer C, Zalinescu C (2015) Regularization of quasi-variational inequalities. Optimization 64:1703–1724
68. Kim TW, Xu HK (2005) Strong convergence of modified Mann iterations. Nonlinear Anal 61:51–60
69. Kirk WA, Shahzad N (2019) Hyperbolic spaces and directional contractions. Bull Math Sci 9:49 pp.
70. Kohlenbach U (2005) Some logical metatheorems with applications in functional analysis. Trans Am Math Soc 357:89–128
71. Kohlenbach U (2008) Applied proof theory: proof interpretations and their use in mathematics. Springer, Berlin
72. Kohlenbach U, Leustean L (2010) Asymptotically nonexpansive mappings in uniformly convex hyperbolic spaces. J Eur Math Soc 12:71–92
73. Konnov IV (2001) Combined relaxation methods for variational inequalities. Springer, Berlin-Heidelberg
74. Kopecka E, Reich S (2004) A note on the von Neumann alternating projections algorithm. J Nonlinear Convex Anal 5:379–386
75. Kopecka E, Reich S (2012) A note on alternating projections in Hilbert space. J Fixed Point Theory Appl 12:41–47
76. Kozlowski WM (2014) An introduction to fixed point theory in modular function spaces. Springer, Cham, pp 159–222
77. Krasnosel'skii MA (1955) Two remarks on the method of successive approximation. Uspehi Mat Nauk 10:123–127
78. Leustean L (2007) A quadratic rate of asymptotic regularity for CAT(0)-spaces. J Math Anal Appl 325:386–399
79. Leustean L (2010) Nonexpansive iterations in uniformly convex W-hyperbolic spaces. In: Leizarowitz A, Mordukhovich BS, Shafrir I, Zaslavski A (eds) Nonlinear analysis and optimization I: nonlinear analysis. American Mathematical Society, pp 193–209
80. Leustean L, Nicolae A (2017) A note on an alternative iterative method for nonexpansive mappings. J Convex Anal 24:501–503
81. Leustean L, Pinto P (2023) Rates of asymptotic regularity for the alternating Halpern-Mann iteration. Optim Lett. https://doi.org/10.1007/s11590-023-02002-y
82. Li J (2016) Iterative fixed point theorems and their applications to ordered variational inequalities on vector lattices. Fixed Point Theory 17:401–411
83. Li J (2019) Several fixed point theorems on partially ordered Banach spaces and applications. J. Nonlinear Convex Anal 20:2095–2108
84. Li L, Xu HK (2021) Further convergence analysis of iterative methods for generalized split feasibility problems in Hilbert spaces. J Nonlinear Convex Anal 22:2575–2589
85. Lopez G, Martin V, Xu HK (2010) Halpern's iteration for nonexpansive mappings. Contemp Math 513:211–230
86. Mann WR (1953) Mean value methods in iteration. Proc Am Math Soc 4:506–510
87. Marino G, Xu HK (2004) Convergence of generalized proximal point algorithms. Commun Pure Appl Anal 3:791–808
88. Mitrovic ZD, Parvaneh V, Mlaiki N, Hussain N, Radenovic S (2020) On some new generalizations of Nadler contraction in b-metric spaces. Cogent Math Stat 7(1):1760189
89. Mitrovic ZD, Radenovic S (2017) The Banach and Reich contractions in bv(s)–metric spaces. J Fixed Point Theory Appl 19:3087–3095
90. Mitrovic ZD, Radenovic S (2019) On Meir-Keeler contraction in Branciari b-metric spaces. Trans A Razmadze Math Inst 173:83–90
91. Mizoguchi N, Takahashi W (1989) Fixed point theorems for multivalued mappings on complete metric spaces. J Math Anal Appl 141:177–188
92. Moudafi A (2000) Viscosity approximation methods for fixed-point problems. J Math Anal Appl 241:46–55
93. Moudafi A (2014) Alternating CQ-algorithms for convex feasibility and split fixed-point problems. J Nonlinear Convex Anal 15:809–818

94. Moudafi A, Gibali A (2018) $l_1 - l_2$ regularization of split feasibility problems. Numer Algorithms 78:739–757
95. Moudafi A, Thakur BS (2014) Solving proximal split feasibility problems without prior knowledge of operator norms. Optim Lett 8:2099–2110
96. Nadler Jr. SB (1969) Multi–valued contraction mappings. Pacific J Math 30:475–488
97. Nicolae A, O'Regan D, Petruşel A (2011) Fixed point theorems for singlevalued and multivalued generalized contractions in metric spaces endowed with a graph. Georgian Math J 18:307–327
98. ODHara JG, Pillay P, Xu HK (2006) Iterative approaches to convex feasibility problems in Banach spaces. Nonlinear Anal 64:2022–2042
99. Qin X, Cho SY, Yao JC (2020) Weak and strong convergence of splitting algorithms in Banach spaces. Optimization 69:243–267
100. Qin X, Petrusel A, Yao JC (2018) CQ iterative algorithms for fixed points of nonexpansive mappings and split feasibility problems in Hilbert spaces. J Nonlinear Convex Anal 19:157–165
101. Qin X, Yao JC (2016) Weak convergence of a Mann-like algorithm for nonexpansive and accretive operators. J Inequal Appl 2016:232
102. Orouji B, Soori E, O'Regan D, Agarwal RP (2021) A strong convergence theorem for a finite family of Bregman demimetric mappings in a Banach space under a new shrinking projection method. J Funct Spaces Art. ID 9551162, 11 pp.
103. Petrusel A, Petrusel G, Yao JC (2016) Coupled fixed point theorems for symmetric contractions in b-metric spaces with applications to operator equation systems. Fixed Point Theory 17:457–475
104. Petrusel A, Petrusel G, Yao JC (2020) Multi-valued graph contraction principle with applications. Optimization 69:1541–1556
105. Petrusel A, Petrusel G, Yao JC (2021) Graph contractions in vector-valued metric spaces and applications. Optimization 70:763–775
106. Petruşel A, Rus IA, Serban MA (2015) Fixed points, fixed sets and iterated multifunction systems for nonself multivalued operators. Set-Valued Var Anal 23:223–237
107. Reich S (1971) Some remarks concerning contraction mappings. Can Math Bull 14:121–124
108. Reich S (1972) Fixed points of contractive functions. Boll Un Mat Ital 5:26–42
109. Reich S (1974) Some fixed point problems. Atri Acad Nuz Lincei 57:194–198
110. Reich S (1978) Approximate selections, best approximations, fixed points, and invariant sets. J Math Anal Appl 62:104–113
111. Reich S (1983) A limit theorem for projections. Linear Multilinear Algorithm 13:281–290
112. Reich S, Tuyen TM (2021) Projection algorithms for solving the split feasibility problem with multiple output sets. J Optim Theory Appl 190:861–878
113. Reich S, Zalas R (2016) A modular string averaging procedure for solving the common fixed point problem for quasi-nonexpansive mappings in Hilbert space. Numer. Algorithms 72:297–323
114. Reich S, Zaslavski AJ (2014) Genericity in nonlinear analysis. Springer, New York
115. Reich S, Zaslavski AJ (2015) Approximate fixed points of nonexpansive set-valued mappings in unbounded sets. J Nonlinear Convex Anal 16:1707–1717
116. Reich S, Zaslavski AJ (2017) Monotone contractive mappings. J Nonlinear Var Anal 1:391–401
117. Reich S, Zaslavski AJ (2018) Contractive mappings on unbounded sets. Set-Valued Var Anal 26:27–47
118. Reich S, Zaslavski AJ (2018) Well-posedness of fixed point problems for monotone nonexpansive mappings. Linear Nonlinear Anal 4:1–8
119. Reich S, Zaslavski AJ (2018) Generic well-posedness of the fixed point problem for monotone nonexpansive mappings. Mathematics almost everywhere. World Scientific, Hackensack
120. Reich S, Zaslavski AJ (2021) Contractive mappings on metric spaces with graphs. Mathematics 9:2774. https://doi.org/10.3390/math9212774

121. Reich S, Zaslavski AJ (2022) Convergence and well-posedness properties of uniformly locally contractive mappings. Topol Methods Nonlinear Anal 61(2):761–773. https://doi.org/10.12775/TMNA.2022.035
122. Reich S, Zaslavski AJ (2023) Fixed point and well-posedness properties of uniformly locally contractive mappings. J Nonlinear Convex Anal 24:2543–2550
123. Reich S, Zaslavski AJ (2023) Convergence of inexact iterates of monotone nonexpansive mappings with summable errors. Axioms 12(1):15. https://doi.org/10.3390/axioms12010015
124. Reich S, Zaslavski AJ (2023) A porosity result regarding uniformly locally contractive mappings. Pure Appl Funct Anal 8:1781–1789
125. Reich S, Zaslavski AJ (2024) Existence of a fixed point and stability results for contractive mappings on metric spaces with graphs. Topol Methods Nonlinear Anal 63:233–244
126. Rockafellar RT (1976) Monotone operators and the proximal point algorithm. SIAM J Control Optim 14:877–898
127. Sabach S, Shtern S (2017) A first order method for solving convex bilevel optimization problems.. SIAM J Optim 27:640–660
128. Sahu DR, Wong NC, Yao JC (2011) A generalized hybrid steepest-descent method for variational inequalities in Banach spaces. Fixed Point Theory Appl 2011:28 p.
129. Sahu DR, Wong NC, Yao JC (2012) A unified hybrid iterative method for solving variational inequalities involving generalized pseudocontractive mappings. SIAM J Control Optim 50:2335–2354
130. Suparatulatorn R, Cholamjiak W, Suantai S (2018) A modified S-iteration process for G-nonexpansive mappings in Banach spaces with graphs. Numer Algorithms 77:479–490
131. Takahashi S, Takahashi W (2007) Viscosity approximation methods for equilibrium problems and fixed point problems in Hilbert spaces. J Math Anal Appl 331:506–515
132. Takahashi W (1970) A convexity in metric space and nonexpansive mappings. I Kodai Math Seminar Rep 22:142–149
133. Takahashi W (2017) The split common fixed point problem and the shrinking projection method for new nonlinear mappings in two Banach spaces. Pure Appl Funct Anal 2:685–699
134. Takahashi W (2018) A general iterative method for split common fixed point problems in Hilbert spaces and applications. Pure Appl Funct Anal 3:349–369
135. Takahashi W, Iiduka H (2008) Weak convergence of a projection algorithm for variational inequalities in a Banach space. J Math Anal Appl 339:668–679
136. Takahashi W, Wen CF, Yao JC (2019) A strong convergence theorem by Halpern type iteration for a finite family of generalized demimetric mappings in a Hilbert space. Pure Appl Funct Anal 4:407–426
137. Takahashi W, Xu HK, Yao JC (2015) Iterative methods for generalized split feasibility problems in Hilbert spaces. Set-Valued Var Anal 23:205–221
138. Takahashi W, Yao JC (2021) Strong convergence theorems under shrinking projection methods for split common fixed point problems in two Banach spaces. J Convex Anal 28:1097–1118
139. Tam MK (2018) Algorithms based on unions of nonexpansive maps. Optim Lett 12:1019–1027
140. Tan B, Qin X, Yao JC (2021) Strong convergence of self-adaptive inertial algorithms for solving split variational inclusion problems with applications. J Sci Comput 8:34 pp.
141. Wang F, Xu HK (2011) Cyclic algorithms for split feasibility problems in Hilbert spaces. Nonlinear Anal 74:4105–4111
142. Wang X, Yang X (2015) On the existence of minimizers of proximity functions for split feasibility problems. J Optim Theory Appl 166:861–888
143. Xu HK (2004) Viscosity approximation methods for nonexpansive mappings. J Math Anal Appl 298:279–291
144. Xu HK (2006) A regularization method for the proximal point algorithm. J Global Optim 36:115–125

145. Xu HK (2010) Iterative methods for the split feasibility problem in infinite-dimensional Hilbert spaces. Inver Probl 26:1–17

146. Xu HK (2010) An alternative regularization method for nonexpansive mappings with applications. In: Mordukhovich BS, Shafrir I, Zaslavski A (eds) Nonlinear analysis and optimization I: nonlinear analysis, volume 513 of contemporary mathematics, pp 239–263. American Mathematical Society, Providence

147. Xu HK, Altwaijry N, Chebbi S (2020) Strong convergence of Mann's iteration process in Banach spaces. Mathematics 8:954. https://doi.org/10.3390/math8060954

148. Xu HK, Cegielski A (2021) The Landweber operator approach to the split equality problem. SIAM J Optim 31:626–652

149. Yao Y, Liou YC, Yao JC (2015) Split common fixed point problem for two quasi-pseudocontractive operators and its algorithm construction. Fixed Point Theory Appl 2015:1–9

150. Zaslavski AJ (2011) Maximal monotone operators and the proximal point algorithm in the presence of computational errors. J Optim Theory Appl 150:20–32

151. Zaslavski AJ (2012) Proximal point algorithm for finding a common zero of a finite family of maximal monotone operators in the presence of computational errors. Nonlinear Analy 75:6071–6087

152. Zaslavski AJ (2016) Approximate solutions of common fixed point problems. Springer optimization and its applications. Springer, New York

153. Zaslavski AJ (2018) Algorithms for solving common fixed point problems. Springer optimization and its applications. Springer, Cham

154. Zaslavski AJ (2019) Approximate fixed points of nonexpansive mappings on hyperbolic spaces. Linear Nonlinear Anal 5:517–524

155. Zaslavski AJ (2021) An iterative method for solving the fixed point problem for a set-valued mapping. Linear Nonlinear Anal 7:355–364